SOCIAL WORK
AND THE
ENVIRONMENT

SOCIAL WORK AND THE ENVIRONMENT

Understanding People and Place

Michael Kim Zapf

Canadian Scholars' Press Inc.

TORONTO

Social Work and the Environment: Understanding People and Place
by Michael Kim Zapf

First published in 2009 by
Canadian Scholars' Press Inc.
425 Adelaide Street West, Suite 200
Toronto, Ontario
M5V 3C1

www.cspi.org

Canadian Scholars' Press Inc. gratefully acknowledges financial support for our publishing
activities from the Government of Canada through the Book Publishing Industry
Development Program (BPIDP) and the Government of Ontario through the Ontario Book
Publishing Tax Credit Program.

Library and Archives Canada Cataloguing in Publication
Zapf, Michael Kim
        Social work and the environment : understanding people and place / Michael
Kim Zapf.
Includes bibliographical references.
ISBN 978-1-55130-357-4

        1. Social service. 2. Human beings—Effect of environment on. 3. Environmental
degradation—Social aspects. 4. Human ecology. I. Title.
HM861.Z36 2009   361.3   C2009-900715-0

Designed and typeset by Em Dash Design

Printed and bound in Canada

Canada

MIX
Paper from
responsible sources
FSC® C004071

To Barbara, Earl, Chris, and Kathy—who created with me the places of my childhood. To Penny and Kevin—who create with me the place I now call home.

CONTENTS

ACKNOWLEDGEMENTS

M any people and many places contributed to the ideas developed for this book. While I have tried to acknowledge all of these contributions, I know the list will be incomplete.

The award of a Killam Resident Fellowship (January to April 2007) provided the opportunity for a focused overview and reflection on my various writings and presentations on social work and the environment over many years. In this way, the Killam Fellowship created the space for completion of the first draft of this manuscript.

Gayla Rogers, Dean of Social Work at the University of Calgary, has encouraged and actively supported my specific work on this book as well as my broader efforts to bring understanding of place into my teaching and writing. Under her guidance, the Faculty of Social Work participates in a multidisciplinary *People and Place* initiative. I hope this book will be of some value to that work.

Susan Silva-Wayne of Canadian Scholars' Press Inc. was most encouraging about the initial proposal. Her optimistic support and guidance of this work through the various stages of review and approval were appreciated.

I also want to acknowledge the contributions of Aboriginal friends, colleagues, healers, teachers, and authors who risked sharing with me some of their experiences and understandings. This group includes Apela Colorado, Phyllis Pruden, Les Jerome, Sharon Big Plume, Betty Bastien, Jeanine Carriere, George Calliou, Rick Lightning, and Michael Hart. I hope I am using my learning with respect.

Twenty-five years ago, I arrived in Toronto from Yukon to begin my doctoral studies. Fortunately, I met Ernie Lightman, a mentor and supervisor who understood place. He said he was willing to work with me as long as I understood that his concept of North was "anything above St. Clair Avenue"! As we worked together, I came to see that it was not necessary to have the same place identifications and experiences. It was necessary to have respect for each other's place identity.

Colleagues in rural and northern practice have had a major influence on my ideas about the physical environment and social work over the years. Roger Delaney and Keith Brownlee of Lakehead University have promoted knowledge-building in this area through the *Northern Social Work* series published by the Lakehead Centre for Northern Studies. Canadian writers such as Ken Collier, Glen Schmidt, and Rosemary Clews have been exploring context-based rural and remote practice for some time. Collaborations with Australian colleagues Brian Cheers, Bob Lonne, and Rosemary Green have revealed similar issues encountered in vastly different geographic settings.

I have shared many hours on rural roads throughout Alberta with colleague and friend Ralph Bodor on our way to community meetings or weekend teaching. Many of the thoughts articulated in this book have their origins in those trips. An amazing teacher and thinker, Ralph is still the only person I have met who can steer with his knees while marking papers.

Although several members have been acknowledged individually, I want to express my overall appreciation to the original BSW Access Team at the University of Calgary (Betty Bastien, Ralph Bodor, Jeanine Carriere, and William Pelech). As we were developing a new curriculum inclusive of context and identity, our discussions were exciting, honest, and challenging. My eyes were opened to new experiences and perspectives on what it means to work and teach in place.

During an alcohol-fuelled discussion one New Year's Eve with my next-door neighbour, Marc Rheaume, I was introduced to the concept of *terroir* wines, an intriguing idea I have come to understand as placemaking captured in a bottle. That discussion prodded me to explore further what other disciplines were saying and doing about place.

A number of organizations have hosted conferences where I had the opportunity to present earlier versions of many of the thoughts and arguments that eventually came together in this book. The feedback, challenges, and connections from those events have been valuable. In particular, I want to acknowledge the Alberta College of Social Workers, the Canadian Association of Social Workers, the Canadian Association for Social Work Education, the Canadian Society for Spirituality and Social Work, the Western Regional Science Association, and the American National Institute on Social Work and Human Services in Rural Areas.

Special mention must be made of John Coates' innovative book on social work and ecology. He gave the rest of us a foundation on which to build, as well as the beginnings of a vocabulary in social work that recognized spiritual connections and responsibilities between people and the natural world. His careful review of earlier versions of my manuscript strengthened this book considerably.

My list of acknowledgements must also include some of the places that have held particular meaning for me. From Noelly's Mountain to Chief Mountain, from Raven Bay to Bay Street, from the Dome of the Rock to the SkyDome, and from the outback to the backyard, I have encountered many geographic points filled with significance. Their meanings have certainly influenced my thinking and my writing.

Finally, I want to thank my wife Penny Ford and our son Kevin for their interest, good humour, and patience as I worked on this book. They were willing to see the dining-room table disappear for months under a shifting, living environment made of piles of books, papers, articles, and draft chapters (a bit like academic coral).

All my relations,

Michael Kim Zapf
October 2008

INTRODUCTION

The Environment, Social Work, and Ecological Thinking

The fate of the Earth is now
intimately intertwined with the
fate of the human species.

(CAJETE, 2000, P. 60)

THE ENVIRONMENTAL IMPERATIVE AND SOCIAL WORK

Towards the end of the 20th century, an urgent wake-up call was sounded at the Earth Summit in Rio de Janeiro. The Indigenous People of Turtle Island issued this challenge to the people of Mother Earth: "Wake Up World—this is not some tug-of-war over a real estate deal; this is a matter of life and death, for we are part of the environment" (Sainte-Marie, 1992). Another dramatic warning came the following year in the form of the *World Scientists' Warning to Humanity*, which was signed by more than 1,670 scientists from 71 countries, including 104 Nobel Laureates (Union of Concerned Scientists, 1992):

WARNING—We the undersigned, senior members of the world's scientific community, hereby warn all humanity of what lies ahead. A great change in our stewardship of the earth and the life on it is required, if vast human misery is to be avoided and our global home on this planet is not to be irretrievably mutilated. (p. 1)

James Meadowcroft (2007), Canada Research Chair in Governance for Sustainable Development at Carleton University, has put forward an insightful analysis of our initial efforts as a society to construct an environmental state for the 21st century, much as we constructed the welfare state in the 20th century. He cautioned, however, that "the establishment of the welfare state took between 50 and 80 years" (p. 15). Such a time frame may not be survivable when it comes to dealing with threats to our physical environment. In his best-selling book and Oscar-winning documentary film *An Inconvenient Truth* (Bender, 2006; Gore, 2006), former American Vice-President Al Gore declared the environmental crisis to be a "planetary emergency." Testifying on climate change before the United States Congress on March 21, 2007, Gore cautioned that "We do not have time to play around with this" (quoted in Kluger, 2007, p. 48).

In the western world, we are witnessing a dramatic rise of environmental concerns to the top of the public agenda early in the 21st century. Public-opinion surveys show that fears related to environmental change have joined crime, health-care, security, and economic concerns as central issues to be addressed by the modern state (Laghi, 2007; Neuman, 2007). There is some evidence that environmental awareness and concern may be concentrated in today's youth (MacQueen, 2007). Labels such as the "New Aquarians" (Adams, 2006) or the "Transition Generation" (Martin, 2006) have been applied with optimism to a new generation equipped with ecological understandings, who will be the ones to bring about great changes to ensure our sustainable future. Having "developed a top-of-mind salience the likes of which we've never seen before" (Allen Gregg, quoted in Laghi, 2007, p. A1), environmental issues appear poised to retain their prominence in the public eye for some time to come.

Of the many environmental issues facing humankind at this time (including deforestation and land degradation, increasing levels of toxic chemicals and pollutants, over-consumption of renewable and nonre-newable natural resources, and overall loss of biodiversity), it appears that climate change is the most immediate and obvious threat, the issue that has pushed environmental concerns to the top of the public agenda. As an example of developing awareness of climate change, consider the rapid evolution of Stephen Harper's position on the issue (as documented by Meadowcroft, 2007). As Leader of the Opposition in 2002, Stephen Harper referred to climate change as "based on tentative and contra-dictory scientific evidence" (p. 17) and to the Kyoto Accord as a "job-killing, economy-destroying … socialist scheme to suck money out of wealth-producing nations" (p. 14). By 2006, now Prime Minister Harper spoke of climate change as based on "emerging science" (p. 17). By 2007, Harper had concluded that "the science is clear" (p. 17) and that climate change "is a serious environmental problem that needs immediate action. Canada's decision to do nothing over the past decade was a mistake, and we want to do better" (p. 14). Various levels of government are now visi-bly incorporating environmental concerns into their priorities and bud-gets in response to identified public concerns (Fekete, 2007; Mittelstaedt, Galloway, & Laghi, 2007).

If environmental issues have reached the top of the public's priority list for action, where is the profession of social work in relation to these new and urgent concerns? Human beings may be entering very difficult times with the degradation and potential destruction of our sustaining natural world. Governments are beginning to respond. Society might even be in the initial stages of constructing an environmental state. Does social work have any relevance as humankind faces these serious chal-lenges? As a profession with a long-standing declared focus on "person-in-environment," social work might be expected to play a leadership role in the planning stages of any new environmental state. Yet we have generally been silent on these serious threats to human well-being and continued existence.

Why is that? Just how has the physical environment been perceived and conceptualized at the core and at the margins of the discipline of social work? To what extent have our foundational assessment and intervention strategies incorporated the physical environment? In what ways might our language, our assumptions, and our traditional knowledge-building approaches be limiting our ability to perceive connections between people and the world we inhabit? Is it time (or past time) for social work to move beyond our conventional metaphor of person-in-environment towards a new paradigm, a new understanding of the relationship between people and the physical environment? These are the questions that I have attempted to answer in this book.

AUTHOR AND VOICE

I have deliberately chosen to use the first-person voice for this discussion rather than the more conventional academic third-person voice. Although I have attempted to be comprehensive and fair in presenting the literature, I must acknowledge that I have been selective in deciding which materials to include in each chapter. The patterns identified in this book are those that have become apparent to me through my work, my reflections, and my writing. All of these activities took place in locations that held meaning for me and influenced my work. I do not want a distant third-person voice to obscure the geographic reality and influence of my located experiences.

The use of a first-person voice allows me to take responsibility for the way I have selected, organized, and presented the material in this book. Over many years of teaching and writing, I have also become keenly aware of the power of storytelling. Inclusion of some of my stories in this book similarly requires a first-person voice. Of course, all of this is not meant to suggest that this book is merely anecdotal. I trust the reader will find the arguments well grounded in the relevant literature, with appropriate citations and references. Academic rigour does not negate a first-person voice for presentation.

Since I have chosen a first-person style of presentation, I must make a few comments to introduce my "voice." I hope that such an introduction might help the reader to understand the approach I have taken in this book. Like everyone else, I am a product of my genetic inheritance, my geography, my culture, my experiences, and the circumstances and issues of my time. I am a white, male, Canadian social worker in my late 50s, currently employed in an academic setting at the University of Calgary. I grew up in a military family that moved every three years. During my childhood, I developed a strong attachment to notions of home and the nuclear family, but with no strong ties to a specific place or location; places were interchangeable and transitory.

My first social work job was in metropolitan Toronto, where I became overwhelmed with the enormous scale of the Ontario housing developments where most of my clients lived. In the early 1970s, I moved west to work on my MSW degree at the University of British Columbia. Those were exciting days for a social work student in Vancouver. A social worker was Premier of the province! Progressive policies merged social concerns with regionalization and land use policies. At school, I learned about the new holistic generalist practice approach based on the systems work of Pincus and Minahan (1973). I graduated with an optimistic sense that I had mastered a powerful and universally applicable problem-solving model of social work practice.

Wanting to work in a smaller-scale environment than I had experienced in either downtown Toronto or Vancouver, I moved to a small community in Yukon. There I encountered a practice reality for which I soon realized I was not well prepared. The First Nations people I met in the area were people who knew where they belonged. I do not mean this in the paternalistic sense of keeping to their assigned places in some social status hierarchy. Rather, I refer to a profound connection with the natural environment that held their identity, their past, and their future. There was a sense of belonging, of being part of the land and the land being part of them. I had experienced nothing like this in my transient military childhood or in my early university and employment years in southern Canadian cities.

The models of human behaviour and approaches to assessment that I had learned in my academic social work training were not adequate to incorporate this special relationship with the natural world. I came to understand that many of the practice models I had learned as universal were actually written by white academics in urban settings, and often these models contained hidden biases and assumptions that did not fit well in the North. Primary amongst these hidden assumptions was a disconnection between people and the natural world. Considerable adaptation was necessary for effective northern practice. During these early years in Yukon, I learned much about the people, the land, myself, and my profession.

In the early 1980s, I decided to return to school to work on developing practice models that might better reflect northern realities. Ironically, the only English-language Canadian Ph.D. program in social work at that time was at the University of Toronto. I actually moved from Yukon to downtown Toronto in search of relevant models of practice for the North! Not surprisingly, I discovered that urban social work virtually ignored the natural world. Writings from Aboriginal social work, remote or northern practice, or spirituality and social work were extremely rare in those days. I found a developing literature on rural social work, but this was mostly American in origin and assumed rural peoples to be disadvantaged urbanites who had to overcome obstacles of distance and attitude to receive the same services enjoyed by people in the city. In short, I was searching for the physical environment in social work and I could not find it. Despite commitment to an ecological perspective and a declared practice focus on person-in-environment, the underlying social work theory base and practice models appeared to me to be deficient when it came to an understanding of, respect for, and engagement with the natural environment.

Since the mid 1980s, I have continued my search for the environment in social work in an academic setting, a search which has led me to many diverse encounters and activities. I continue to learn from Aboriginal colleagues and communities at home and from international colleagues who have been engaged in similar work in such regions as rural America, outback Australia, and northern Scandinavia. I have attempted to understand how other disciplines conceptualize the relationship between people and the natural environment by attending and presenting my work at conferences of geographers, regional scientists, and educators. My curriculum development work at the University of Calgary has resulted in credit courses with a strong environmental component (for example: Rural & Northern Practice; Generalist Practice in Context; People and Place). To this point, my work has been largely self-directed, motivated primarily by my frustration at the neglect of the natural environment in the social work literature, and my interest in the perspectives offered from Aboriginal culture and other disciplines. My efforts have resulted in a patchwork of journal articles and book chapters, all with strong connections to the physical environment. Given the current public priority assigned to environmental issues and the political changes that have begun, I believe it is now time to consolidate these understandings of social work and the environment into one volume.

While I first became aware of connections between social work and the physical environment during my practice in rural, remote, and Aboriginal communities, the arguments presented in this book are not intended only for practitioners or students in those regions. This is not simply a rural social work book. The mainstream, urban-based profession faces monumental challenges of relevance in responding to society's environmental concerns. My analysis may have originated in my rural/remote experiences, but the message is directed to the larger profession. I hope this book will help to bring environmental concerns into the practice, policy, research, and education domains of our overall profession.

Although I have identified my voice as Canadian, this book is intended for an international audience. The literature base for the discussions that follow includes material from around the world (with an admitted emphasis on Canadian, American, and Australian sources). Ecological concerns transcend conventional national boundaries. We are becoming aware of new threats to all of humankind on the planet. Our initial understandings may be grounded in localized experiences, but effective responses must eventually be globally coordinated.

ORGANIZATION OF THE BOOK

This book begins with an examination of the historical and present treatment of the natural environment within mainstream social work. Using a metaphor from magic, Chapter 1 makes a case that the broad notion of environment has generally been transformed into the much narrower social environment. With no rationale or explanation offered, *person-in-environment* has become *person-in-social environment*, thereby excluding most consideration of the natural world or of clients' physical environments. Chapter 2 examines passing references to, and limited applications of, the physical environment in the recent mainstream social work literature. Rare efforts to fully incorporate the physical environment into our assessment and intervention strategies are highlighted in Chapter 3.

The next chapters of the book look to specializations at the fringe of the social work profession for alternative understandings of the physical environment. Chapter 4 presents the importance of context and place in rural and northern social work. Chapter 5 examines spiritual connections between people and the natural world, as expressed in the developing literature on spirituality and social work, deep ecology, and Aboriginal social work. International social work is considered in Chapter 6, with an emphasis on notions of sustainable development.

Concepts that may prove useful for social work can be found in a number of other disciplines that also explore person–environment relationships. Chapter 7 offers perspectives on environment and place that are found in such diverse modalities as painting, film, music, viticulture, sociology, psychology, environmental design, geography, and education.

Building on the material from social work and related disciplines, the concluding chapters of the book consider where we might go from here. Chapter 8 identifies some of the difficulties and limitations we face in social work when trying to express new conceptualizations of the relationship between people and the natural world. Chapter 9 concludes with an argument for replacing our familiar and limiting notion of person-in-environment with a new understanding of people as place, which may be useful for the immediate and long-term work ahead.

Throughout this book, I make frequent use of two terms that need clarification because they can have varied and confusing meanings. The first of these potentially difficult terms is "physical environment." I use the term to distinguish the physical world from the social environment, the economic environment, the political environment, or any other socially constructed context. The physical environment generally includes both the natural environment and the built environment. The second term that carries considerable baggage is "mainstream," as in "the mainstream social work literature." Somewhere along the line, "mainstream" acquired a negative connotation of meaning ultra-conservative, unimaginative, or simplistic. This is not the sense in which I use the term in this book. The mainstream social work literature is the body of knowledge found in the core textbooks and most widely circulated journals of the profession. There is no negative judgment implied by reference to mainstream social work or its knowledge base. I may critique elements of that knowledge base for its neglect of the physical environment, but not because it is mainstream.

Thinking Ecologically

Fears of the implications of climate change and sudden renewed public interest in the natural environment have created a context where society may be facing a fundamental shift in values and approaches towards living on and with this planet. If indeed humankind is venturing into a new era of environmental citizenship and the possible construction of an environmental state, then it is important for those of us in the social work profession to be clear about the knowledge and perspectives that we bring to this task.

Morito (2002) clarified an important distinction between thinking about ecology and thinking ecologically. When we simply think about ecology, we treat environmental concerns as a new and distinct subject area with its own issues, jargon, and approaches to building knowledge that are disconnected from other human concerns. Environmental problems are best left to the specialists: the environmentalists. On the other hand, "When we begin to think ecologically, we begin to understand ourselves, perceive, judge, analyze, formulate concepts and assume responsibilities differently. The underlying thought process is fundamentally different from that of the now-present dominant mode of thought" (Morito, 2002, p. 9).

Ecological issues cannot be relegated to one separate discipline with responsibility for the environment. Ecological thinking is a process, a world view, a set of principles, and an awareness that must affect all approaches to enquiry and practice if we are to survive. Following Morito's distinction, I have not written a book about ecology from a social work perspective. Rather, I have written a book about the importance of our profession learning to think and act ecologically if we are to be relevant in addressing the serious environmental concerns facing humankind in the 21st century.

CHAPTER ONE

SMOKE AND MIRRORS

How the Environment Became the Social Environment

> *Smoke and mirrors* is a metaphor for a deceptive,
> fraudulent or insubstantial explanation or description.
> The source of the name is based on magicians'
> illusions, where magicians use smoke and mirrors
> to accomplish illusions such as making objects
> disappear, when they really don't disappear at all.
>
> (WIKIPEDIA, 2007)

Deception? Insubstantial explanations? Smoke and mirrors? What does any of this have to do with social work and the environment? In this chapter, I show how the knowledge base of mainstream social work practice has perpetrated just such a deception, or at least failed to offer a substantial explanation, as the *environment* has been changed before our very eyes to—Presto!—the *social environment*. Of course, the physical environment has not really disappeared at all. Not only is it still here, but we are becoming ever more aware of its presence and the implications of our profound interrelationship. Yet mainstream social work is having a difficult time in recognizing, let alone directly confronting, environmental challenges. Like an audience at a magic show, we have been duped into accepting a limited set of assumptions as reality, whereby the physical environment has disappeared or been rendered irrelevant.

How did this come about? How was the illusion accomplished? Can we see through the smoke and mirrors? To begin to answer these questions, I start with my own first encounters with the illusion.

THE EFFECT: PERSONAL OBSERVATIONS OF THE VANISHED ENVIRONMENT

In the early 1980s, I left my practice in Yukon to begin work on my doctoral degree at an urban university in southern Canada. Excited at the prospect of building relevant northern practice models and applications, I was eager to apply the developing generalist practice knowledge base to my experiences from the North. My first encounters with the new literature, however, proved to be both confusing and disappointing.

I was initially encouraged when I came across the first edition of Turner and Turner's (1981) *Canadian Social Welfare,* which opened with an evocative photograph of Commissioner Stu Hodgson of the Northwest Territories standing beside an airplane on a snow-covered runway with his arms around two Inuit children, all smiling through their fur-lined parka hoods. The only photo in the book, this image apparently set the tone for a discussion of social welfare in the Canadian context. I was soon disappointed, however, as the rest of the book virtually ignored the physical context of social service delivery. The picture had been smoke and mirrors.

In 1984, Canada was privileged to host the Eighth International Symposium of the International Federation of Social Workers (IFSW). A special issue of the Canadian Association of Social Workers (CASW) journal *The Social Worker* was assembled "to reflect Canadian realities" and "provide a state of the art summary of the profession" (Torjman, 1984, p. 3) in this country for foreign delegates. The first major issue identified for Canadian social welfare was our "geographic diversity and rural/urban imbalance" (Drover, 1984, p. 6). I was initially encouraged to see defining aspects of our natural environment identified as crucial to our understanding of social work in Canada, yet I was once again disappointed to discover no further mention of the physical environment throughout the journal's entire overview of Canadian practice realities. The introduction had been smoke and mirrors.

Later I came across Yelaja's (1985) book *An Introduction to Social Work Practice in Canada*, in which the ecological metaphor was presented as a major influence on Canadian social work with its emphasis on "the reciprocal relationships between the individual and the environment and the continuous adaptation of both person and environment to each other" (p. 29). Great! This notion fit well with my lived experiences of northern peoples adapting to the northern environment. Yet Yelaja immediately went on to say that, from this ecological perspective, "human growth and development constantly change in relation to the social environment—and the social environment changes in response to human factors" (p. 29). Whoa! What happened? Within two sentences, the "environment" had become the "social environment" with no rationale or explanation offered, not even an acknowledgment of the transition. Smoke and mirrors again—the physical environment had disappeared, effectively written out of the ecological equation.

What was going on? Realizing that I might be particularly sensitive to the influence of the physical environment from my recent experiences living and working in Yukon, I was still convinced that I was beginning to see a real but unacknowledged pattern in the social work literature of the time. The "environment" would be introduced as a defining or foundational aspect of social work, and then either dropped entirely or transformed into the "social environment" without explanation.

From my own amateur magic experience (performing as The Great Takhini!), I knew that when I bought a magic trick, the accompanying instructions generally consisted of two sections: the effect and the method. The effect was what the audience saw. The method was how the magician actually did the trick—the misdirection, apparatus, or gimmick (the unnoticed secret) that served to create the effect for the audience.

With regard to social work and the environment, I knew what the effect was from my frustrating encounters with the literature. The physical environment had vanished, with only the social environment left in its place. As a member of the social work audience, this is what I saw. Yet I was not clear on the method. How was this happening? I had doubts about any grand conspiracy among social work authors to perpetrate a deliberate and secret manipulation against the physical environment. To return to the idiom of "smoke and mirrors," I suspected a case of "insubstantial explanation" rather than outright fraud. Was there something at the core of mainstream social work that (mis)directed our attention away from the physical environment? I began to find answers when I turned to the roots of our ecological perspective.

BEGINNINGS: LIGHTING THE MATCH

In her foundational book *What is Social Work?*, Mary Richmond (1922) made a distinction between the social and physical environments as influences on human behaviour. Yet she concluded that the physical environment "becomes part of the social environment" to the extent that it "frequently has its social aspects" (p. 99). The pattern that I first encountered in the 1980s apparently had its roots in the early days of our profession some 60 years earlier! Richmond had acknowledged the physical environment as an important contextual consideration for social work, but perceived its importance only in terms of its social aspects. From the outset, the profession of social work may have been more comfortable using social science lenses to view the environment rather than perspectives from the physical or natural sciences.

I assumed that this exclusive focus on social aspects of the environment might effectively be challenged when the profession later adopted an ecological perspective from the natural sciences, but I was wrong. Instead, the ecological perspective was distorted to reaffirm the profession's emphasis on the social environment. Consider Gordon's (1969) work that called attention to social work's newly declared "simultaneous dual focus on organism and environment" (p. 6). Here is a statement of the ecological perspective perhaps in its purest form, but Gordon immediately went on to declare his assumption that the organism would be "interpreted by psychological theory" while the environment could be "interpreted by sociological and economic theory" (p. 6). Similar to Richmond's work, here was another clear instruction to understand the physical environment in social terms using perspectives from the social science disciplines of sociology and economics. Gordon (1981) later asserted that "the ultimate goal of social work is to bring about a balance between the realities of a person's capabilities and a person's social situation" (p. 136), with no mention of the physical environment at all.

Developing the functional systems perspective, Pincus and Minahan (1973) proposed four basic systems for social work practice: the *change agent system* (worker and agency), the *client system* (person[s] asking for help), the *target system* (person[s] who have to change in order to accomplish the goals), and the *action system* (everyone directly involved in the change effort). All of these systems are social. From this systems perspective, "the focus of social work practice is on the interactions between people and systems in the social environment" (Pincus & Minahan, 1973, p. 3). The goal of this approach to social work appeared to be the restoration of balance or equilibrium within immediate social systems when there had been some disruption. Broader social change or institutional transformations were seen to be beyond the scope of social work, let alone any consideration of the physical environment.

THE 1980s AND 1990s: STILL SMOKIN'

I began this chapter with specific examples of the neglect of the physical environment in the social work literature that I encountered in the

1980s. There are many more. Consider a chapter with the promising title of "Environmental Modification" (Grinnell, Kyte, & Bostwick, 1981). This chapter begins with a direct discussion of the meanings of "environment," some of which clearly include the physical environment. To my surprise, and again with no explanation, the authors then present their practice model with assumptions, concepts, and roles that relate to modification of only the social environment. Later in the decade, Allen-Meares and Lane (1987) added to my confusion with the language they used to clarify the ecosystems approach as the foundation for grounding practice in theory: "From its beginnings, social work's mission has been to improve the interaction between persons and their natural social environment" (p. 515). What on earth was a "natural social environment"? Would it be anything like the social natural environment perceived earlier by Richmond (1922)? The ambiguous and confusing language did little to clarify my understanding of the relative importance of the physical environment and its relationship with human activity.

Throughout the 1980s, the social work literature neglected the physical environment either by omission or by subordination to the social environment. For the most part, I found that this pattern of neglect continued through the mainstream social work literature of the 1990s. Many sources would begin with a statement of commitment to an ecological perspective that appeared inclusive of the physical environment, only to reverse that stand immediately or later in their work. Selected examples from social work textbooks of the 1990s serve to illustrate the continuation of this curious and frustrating pattern.

Hoffman and Sallee (1994) introduced their model of generalist social work using a metaphor of bridge-building. They explained how practitioners "must understand people as they interact with their environments, as they are affected by their environments, and as they seek to have an impact upon their environments" (p. 3). However, they soon retreated to a much narrower focus for generalist practice "at the point where the person and social environment interact" (p. 10). Tolson, Reid, and Garvin (1994) similarly presented generalist practice as rooted in a concern for "individual-in-environment and the transactions between the two," but immediately defined the environment in terms of "social conditions" (p. 2).

In a book affirming the ecosystems perspective as a foundation of social work practice, Meyer (1995) reminded us that "since the beginning of the profession, practice has focused on the person and the environment" (p. 16). Declaring this person-and-environment perspective to be a "psychosocial focus," she then relegated the environment itself to marginal status as an "outer aspect" of any case (p. 16). The physical environment was conspicuously absent from Meyer's list of "elements of practice emphasized in the ecosystems perspective" (p. 17).

Hancock (1997) substituted "situation" for "environment." He focused on "person-and-situation as a single entity" but defined "situation" exclusively in social terms, including "whole family context, ... workplace, and the client system's membership and status within the local and the wider community" (p. 7). Kirst-Ashman and Hull (1997) asserted that "intervening at the macro level requires a solid recognition of the community as an ecological system" (p. 273), but they defined community as part of the social environment. Only four sentences in their 576-page macro-practice book were devoted to "environmental situations [that] can also create stress" (p. 499), and these had mostly to do with agency conditions causing stress for the worker.

Hardcastle, Wenocur, and Powers (1997) presented several definitions of community that included aspects of geographic space and physical structures, observing that "connection to a territorial base is frequent" (p. 97). The definition of community adopted for their book included a potential dimension labelled a "functional spatial unit" (p. 97). If community often involves a spatial component—a geographic location, an identifiable place in the physical environment—why then was the unifying theme of community practice presented in their macro-practice textbook as "individual lives entwined with the social environment" (p. 415)? Surely individuals are also intertwined with the physical environment, a position that might be assumed from the authors' chosen definition of community. Once again, the physical environment disappeared without explanation.

A case study by Maluccio, Washitz, and Libassi (1999) began with a declared ecological perspective, including attention to the "client's physical or social environment" (p. 32), but then added competence-centred practice to promote an overall approach with the awkward label of "ecologically oriented competence-centered social work" (p. 32). The physical environment may have been included at the outset, but the case featured an "emphasis on restructuring and enriching the client's environment by identifying and mobilizing resources in the clients and their social networks" (p. 33). Once again, the physical environment disappeared.

By the late 1990s, Johnson (1998) was arguing that social workers need to have knowledge of environmental factors for their work, factors that included "social, economic, and geographical and climactic conditions that are a part of the immediate surroundings of the individual" (p. 9). Introducing "environmental manipulation" as a potential strategy for social work, Johnson lamented that the knowledge base for such work "has remained in the realm of practice wisdom or common sense" (p. 368). What else could be expected? Maybe the reason that social work approaches to the physical environment remained underdeveloped in the 1980s and 1990s was because we had been defining the physical environment as external to our primary focus for eight decades!

THE 2000s: ILLUSION REAFFIRMED

Neglect of the physical environment continues in the mainstream social work literature of the 2000s. Social work's knowledge base integrates material from other disciplines, including psychology, sociology, political science, economics, biology, history, education, psychiatry, and anthropology (Chappell, 2006; Kirst-Ashman, 2007; Zapf, 2005b; Zastrow, 2004), most of which consider the individual or society as the subject and the natural world as the background. In spite of social work's declared dual focus on person-and-environment, core theories and models of intervention have continued to emphasize the person over the environment, with little or no concern for the physical environment at all. Numerous examples from the literature support this assertion.

McKay (2002) declared the defining feature of social work as "its dual focus on person and environment," then went on to observe that, for mainstream practitioners, "the importance of attempting to change the social environment is implicit in their emphasis on person–environment transactions" (p. 21). The environment was once again transformed into the social environment without explanation. Miley, O'Melia, and DuBois (2004) similarly set out a promising view of transactions between people and their environments: "People affect their environments and, likewise, the social and physical environment affects people" (p. 34). Yet on the very same page they reaffirmed "social work's focus on social functioning," which they presented as the balance between coping efforts and the demands of the "social environment." They went on to acknowledge the importance of "contextual influences" on individual development and behaviour (p. 35), but the crucial contexts listed were all social, with no mention at all of the physical environment as a context.

The ecosystems model put forward by Morales and Sheafor (2004) included a level of analysis labelled as "environmental-structural factors," but these were defined as "political, economic, and social forces in the social environment" (p. 232). In the preface to their book of generalist-practice case studies, Rivas and Hull Jr. (2004) explained that they had purposely selected cases to present a diversity of practice skills, client systems, fields of practice, populations at risk, and cultural backgrounds. No mention was made of the physical or natural environment. Fifteen "general questions to assist in case analysis" were offered (p. xxii), but none dealt with the physical environment or context. How will we ever see the potential physical resources available for our work if we are trained to ignore them in our assessments and analyses of case situations?

Compton, Galaway, and Cournoyer (2005) asserted that "we should not dichotomize the individual and the environment. They are not separate entities. The person is part of the environment and the environment is part of the person" (p. 40). Yet the very next sentence clarified that "Indeed, the structures and cultures of society emerge in the self and life of each person just as each person affects the environment" (p. 40). Poulin (2005) also declared the heart of the ecosystems perspective to be "the *person-in-environment concept*, which views individuals and their environments as an interrelated whole" and then went on, three sentences later, to explain how "the relationship between individuals and their social environment is reciprocal…. The social environment influences individuals' perceptions of themselves and their interactions with others. Individuals, in turn, influence their social environments" (p. 27). It would appear that the environment of which we are a part is a social one that involves influences of culture, perception, and social structures, but not the natural world.

Carniol (2005) wanted to push past *conservative* approaches to social welfare (ecological and systems perspectives; focus on interactions and imbalances between individuals and social systems; goals related to maximizing human growth and well-being) to advocate for *progressive* approaches (structural, anti-oppressive, and critical perspectives; consideration of systemic oppression; goals related to emancipation). Mullaly (2002) also emphasized goals of social transformation and social justice using the language of anti-oppressive social work. These progressive structural approaches, however, were not applied to issues relating to the physical environment. There was no discussion of environmental justice or systemic oppression of the environment for economic gain.

Heinonen and Spearman (2006) explained that "the primary focus of social work should not be on psychological forces, the environment, or the social structure, but on the interface or relationship between the per-

34

son and the social environment" (p. 182). This was an interesting and efficient instance of the familiar smoke and mirrors switch. In a single sentence, the triad of person, environment, and social structure became the duality of person and social environment.

Hick (2006) concurred that a person-in-environment approach is what continues to distinguish social work from other helping professions. He defined these "environments" to include "immediate family, ... friends, neighbourhoods, schools, religious groups, laws and legislation, other agencies or organizations, places of employment and the economic system" (p. 20). Although the term "social environment" is not used, all of the elements comprising Hick's "environments" are social. Chappell (2006) also identified only social forces in the environment that influence behaviour. "To understand a client's *environment*, a social worker examines how a client's personal functioning may be affected by the general culture, economy, political climate, and other external forces" (p. 151). Are there no physical, natural, or spiritual forces that influence behaviour or sit as potential resources in the client's environment?

In his discussion of untapped resources that are potentially available in a client's environment, Saleebey (2007) used the language of nature to present the environment as "a potentially lush topography of resources and possibilities" (p. 19), but in the very next sentence presented those lush resources in purely social terms: "In every environment, there are individuals, associations, groups, and institutions who have something to give, something that others may desperately need" (p. 19). Later in his book, Saleebey returned to this theme of how an environment can be "rich with resources," but listed the resources as "people, institutions, associations, [and] families" (p. 89). Why was the literal "lush topography" of the natural surroundings not considered as a potential resource?

Kirst-Ashman (2007) made reference to "harmony with nature" in her generalist-practice textbook, but only as a "cultural value" of Native Americans (p. 68), a "conceptual theme that characterizes American Indian cultures" (p. 74). Such balance or harmony with nature was limited to a characteristic of Aboriginal peoples to be considered when providing services to only that population. Harmony with nature was not considered as a theme or value of mainstream society or of the social work profession.

In keeping with the magic metaphor from the beginning of this chapter, I conclude this list of examples of the disappearance of the physical environment from the social work literature of the 2000s with what I call the Grand Illusion. This is arguably the most blatant example of smoke and mirrors, the purest instance of an unexplained shift from the broad notion of environment to the narrower consideration of only its social components. The transformation is accomplished with astonishing speed and efficiency in the subject index at the back of Hull Jr. and Kirst-Ashman's (2004) text *The Generalist Model of Human Services Practice*. Under the entry for "Environment" (p. 483), it simply says "See Social Environment"! No explanation. The physical environment is gone in only three words. Poof! Smoke and mirrors!

VARIATIONS ON THE ILLUSION

Not all writers in the social work field have caused the physical environment to vanish through simple substitution of the social environment. There have been variations. Some have challenged the very centrality of the ecosystems perspective to the profession. Others have accepted the perspective, but limited related practice skills to mechanisms of social coping. Still others have relegated ecological concerns to a lower status as a subset or subordinate component of the profession's overall approach.

Following a content analysis of some 8,000 abstracts from 40 social work journals, Rogge and Cox (2001) found that only 7% of the abstracts contained the terms "person-in-environment," "ecological," or "environment" (p. 55). They concluded that "the person-in-environment perspective has played a small role in core social work journals over the last decade" (p. 66). Acknowledging the historical importance of the ecological approach as fundamental to generalist practice, Heinonen and Spearman (2006) cautioned that it is not an explanatory theory but rather a "loose connection of ideas, views, and concepts" providing a "framework for assessment" (p. 184). Perhaps such a "loose connection" cannot be expected to account for the physical environment and its impact on human activity and growth. Cossom (2002) suggested that the ecological perspective in social work was never intended to be applied to the physical environment at all; it was merely a model for understanding social relationships, "built on biological studies of interrelationships between organisms and their environment in attempts to formulate a better understanding of the complexities of people in relation to their social environments" (p. 395). At least he offered a rationale for the transition from physical environment to social environment.

Kirst-Ashman (2007) identified "coping" as a key concept that has been adapted from the ecological approach for use in social work. Coping is defined as "the struggle to adjust to environmental conditions and overcome problems" (p. 25). In my opinion, this constant struggle to adapt—to cope—characterizes all life forms from an ecological perspective. It seems to me that this *is* the ecological perspective in its purest form. If adjusting, coping, and prevailing in the environment are central to ecology, why then did we take only the social environment as our focus in social work?

According to Kirst-Ashman (2007):

> The *social environment* includes the conditions, circumstances, and interactions that encompass human beings. Individuals must have effective interactions with their environment to survive and thrive. The social environment involves the type of home a person lives in, the type of work a person does, the amount of money that is available, and the laws and social rules people live by. The social environment also includes the individuals, groups, organizations, and systems

with which a person comes into contact, such as family, friends, work groups, and governments. (p. 25)

There is a hint of leakage here as "the type of home a person lives in" introduces aspects of the built environment. But for the most part, coping behaviour, central to the ecological perspective, is considered as effective interaction and coping with social systems. Coping with the natural or built environment is apparently not important for human beings to "survive and thrive" (p. 25). Try telling that to victims of a natural disaster, incarcerated prisoners, prairie farmers, or Inuit facing changing ice conditions.

Johnson and Yanca (2007) presented the ecosystems approach as "a subset of the ecological perspective [that] involves all the systems in a person-in-environment approach, including the physical environment" (p. 13). This is a curious notion of subset. Generally speaking, a subset is smaller than the original set. A subset consists of components selected according to some criteria from the original set. Why would an ecosystems approach be considered a subset if it includes all the systems of the original ecological perspective *plus* the physical environment? This makes no logical sense in terms of size or number of elements. Perhaps the label "subset" is applied to denote subordination. Maybe any approach that includes the physical environment is deemed to be less than the mainstream approach geared to the person and his or her social environment. Later in the book, Johnson and Yanca (2007) advocated understanding the environment as an ecosystem and suggested that generalist social workers seek "to understand the feelings, thoughts, and actions of people; the human systems in the environment; and of the transactions of people and systems in the environment" (pp. 128–129). It would appear that the ecosystems approach, as defined at the outset of the book to include the physical environment, has now been redefined in terms of only social systems and transactions. Once again, the physical environment has been lost (but at least it took more than 100 pages to do it this time!).

Using the smoke and mirrors metaphor, this chapter has offered evidence from the mainstream social work literature of a continuing process whereby the environment has been transformed into the social environment, with the physical environment disappearing altogether. No adequate explanation has been offered for this change, and the switch is often unnoticed. Fortunately, not all social work authors have left the physical environment behind. A minority have taken the step of declaring the physical environment to be an integral component of their world view and foundation for practice. Sadly, most of these efforts have been undercut by less than full support for the environment in subsequent applications. In some cases, ecological language has been used only as window dressing for conventional approaches. In others, the natural and built environments are left out of assessment tools and practice models. Some have limited considerations of the physical environment to very narrow applications such as agency furniture arrangement. Environmental concerns have also been presented as the domain of environmentalists and not social workers. The next chapter explores these partial attempts at inclusion of the physical environment.

CHAPTER TWO

FALSE STARTS AND SPUTTERINGS

Partial Efforts to Include the Physical Environment

> Most people now live in the human-created
> environment of big cities where it's easy to
> believe the illusion that we have escaped our
> biological dependence on the natural world.
>
> (SUZUKI, 1999, P. 45)

I n the previous chapter, I identified a pattern in the social work literature whereby our profession's declared focus on person and environment has quickly and repeatedly become transformed into a consideration of person and social environment. In the many examples given, the
physical environment has been relegated to marginal status or, most often,
ignored completely. While I observe that this pattern of neglect has been
pervasive in the social work literature, I must also acknowledge that some
authors have called for the physical environment to have a central place in
our theory and practice approaches. In this chapter, I examine works that
begin with a strong general commitment to the physical environment, and
then undercut this position through limited application or practice mod-

els that do not include the physical environment. While Chapter 1 presented a pattern of neglect and omission, the efforts examined here might better be described as a series of false starts or sputterings.

Much like a vehicle that refuses to start on a cold Canadian winter morning, the literature has turned over a few times but refused to fully engage with the physical environment, to bring it to life. But at least the battery isn't dead! The engine may yet catch and allow the profession to move. Maybe the necessary first step is to simply acknowledge the importance of the physical environment, even if most of these accounts eventually disappoint by failing to maintain its centrality in the subsequent models and approaches presented.

THE PHYSICAL ENVIRONMENT AS WINDOW DRESSING

There was a time in the history of social work when we adopted medical phraseology to describe our work. Influenced by the dominant medical model of the time, we engaged in *social diagnosis* in order to *treat* clients who *suffered* from major *social ills*. As awareness of our deteriorating physical environment begins to occupy more and more of the public centre-stage, I find scant evidence that the language of our helping profession is adapting to incorporate this new emphasis. Alle-Corliss and Alle-Corliss (1999) spoke of "social pollutants" (p. 27) that must be tackled to preserve healthy social environments. Morales and Sheafor (2004) referred to "noxious social conditions that impede the mutually beneficial interaction between persons and their environment" (p. 231). Mention of pollution and noxious elements may initially suggest a concern with the physical environment, but the terms appear to be window dressing as these authors only concerned themselves with social forces and conditions throughout the body of their works.

Garvin and Seabury (1997) sought to promote notions of competence and social justice for social workers. Although focused on interpersonal practice, their book included a very brief reference to stressful

conditions in the physical environment "such as pollution, lack of material resources, or dangers," conditions that were said to be "often present
in the physical environment of poor people" (p. 185). I am pleased to see
that the physical environment was at least mentioned, but it seems to me
that this kind of brief mention, a passing reference, is more of a dismissal
than a valid effort at inclusion. Why were these environmental concerns
excluded from the assessment approaches and intervention methods discussed throughout the book? I also find it a curious notion to suggest that
environmental concerns might be relevant to social work because they are
common in the milieu of "poor people." If this is true, then should not
these same environmental concerns constitute an integral component of
our analyses of oppression and our perceptions of injustice? The one-sen-
tence reference to the physical environment in this book raises far more
questions than it answers.

Other examples of superficial passing references to the physical environment can be found in the social work literature. Hepworth, Rooney,
and Larsen (1997) devoted a single sentence of their text to mention of
environmental "hazards and pollution" (p. 266), with a similar effect of dismissal. Mattaini, Lowery, and Meyer (2002) produced a book advocating
an "ecosystems perspective" for social work, an approach said to emphasize "the connectedness among case elements, consistent with contemporary physical, behavioural, and ecological science" (p. 20). The connections
with physical and ecological science strongly suggest a place for the natural world in their model, but the ecosystem scan-assessment guide for
work with individuals (Mattaini, 2002) addresses only social content, and
the chapter on practice with organizations (Hanson, 2002) similarly limits
the environmental context to social components of the agency environment. In a 149-page section on the implementation and evaluation phase
of social work practice, Miley et al. (2004) devoted five sentences to the
issues of "designing environments" (p. 352). Passing reference was made

to the influence of "the physical design of space" on relationship development and social interaction, but this was left undeveloped.

Compton et al. (2005) similarly passed over the physical environment with a general direction that, in addition to all the facets of the social environment illustrated in their diagrams, social workers must also consider "the natural world, such as geography or climate, and the constructed world, such as the shelters or roads we build" (p. 48). After all of the components of the social environment have been identified and thoroughly examined for their importance to social work practice, the natural world and the constructed world are simply dismissed with a nod. We are told that many social workers "regularly help clients to change environmental conditions that affect problems of concern" (p. 53), but this is dropped immediately and left completely undeveloped. What environmental conditions affect client problems? How? Are they resources or liabilities? How do social workers tackle environmental conditions with clients? Toward what ends? According to what values? Using what methods? None of these questions was addressed as it was for the social environment throughout the book.

These false starts can be frustrating. On the one hand, I am pleased to see the physical environment at least mentioned in these practice texts. On the other hand, I perceive a definite sense of exclusion or dismissal when the physical environment is presented as window dressing, an afterthought, an undeveloped add-on.

THE PHYSICAL ENVIRONMENT LIMITED TO AGENCY OR CLIENT ENVIRONS

Some social work authors have declared the importance of the physical environment, but then limited or constrained this notion through a very narrow focus of application. One group has looked only at the immediate physical environment of the social agency, while others have looked only at the client's immediate physical environment.

After affirming that the physical environment sends messages to people and influences relationships and behaviour in particular settings, Resnick and Jaffee (1982) for some reason chose to focus only on "the immediate exteriors and interiors of the physical structures in which social services are delivered and the nature and use of the space within these structures" (p. 354). Arguing for more attention to be directed towards the physical environments where people play out their lives, Saleebey (2004) similarly concluded that "it is important for social workers to be aware of the conditions of the environment of the agency in which they work" (p. 14). Corey and Corey (2007) acknowledged "the physical aspects of the work setting" (p. 357) as a possible stressor for social workers, but inexplicably proceeded to define such physical aspects only in social terms (workload, office politics, financial cutbacks, and relationships with colleagues and supervisors).

Neugeboren (1996) took this same curious turn in a book entitled *Environmental Practice in the Human Services*. Despite the broad notion of environmental concern suggested by the promising title, the book quickly shifts emphasis to "creating positive environments for service consumers" (p. 18) with an emphasis on the social environment. In fact, the entire book includes only one paragraph (p. 251) that deals directly with the physical environment, and this is completely focused on an agency's physical space (with mention made of lighting patterns, non-skid surfaces, safety features, and corridor length).

Of those authors who have considered the immediate physical environment of the client, Hepworth et al. (1997) made passing reference to client living conditions, including space, living arrangements, sanitation, privacy, heat, and nutrition (p. 266). Hull Jr. and Kirst-Ashman (2004) similarly identified "the types of homes people live in" (p. 9) as an area of concern for generalist practitioners, but they included residence as part of the social environment. Miley et al. (2004) briefly introduced notions of privacy, personal space, lighting, and ventilation as "important factors to assess" (p. 257), but left these concerns undeveloped. Shulman (2006) agreed that features of the physical environment are often overlooked in social work practice, but the examples given were limited to the worker's office furniture and the client's construction of personal space at home.

Saleebey (2004) wanted social workers to pay attention to the "near environments of clients" (p. 14), the living spaces, the neighbourhoods, "the immediate, varied and shifting environs where people live day to day" (p. 15). This attention to the physical environment was immediately undercut, however, as he declared that "this requires of us putting the social ahead of the individual, or at least putting the social on equal footing with the individual in assessment and in practice" (p. 15). What happened to the living space, the near environment, the varied environs where daily life unfolds? After such a promising beginning, social environment took over once again.

Diagrams and the Disappearing Physical Environment

There are several instances in the literature where authors have declared inclusion of the physical environment in their statements of world view or practice foundations, and then inexplicably left the physical environment out of their diagrammed models of practice.

Netting, Kettner, and McMurtry (1993) incorporated the physical environment into part of the larger community within which clients live and function. Space, geographic boundaries, population distribution, and physical barriers were all addressed in the questions to be asked during community assessment. However, the graphic representation of "the individual within the community," a diagram of overlapping circles (p. 71), does not include any of these features of the physical environment.

Hepworth et al. (1997) observed a new trend of attention "accorded to environmental factors and to understanding ways in which people interact with their environments" (p. 17). Through this interaction, people do more than simply react to their surroundings; "rather, they act on their environments, thereby shaping the responses of other people, groups, institutions, and even the physical environment" (pp. 17–18). Here is the physical environment front and centre, engaged with people in interaction and mutual influence. Why then are the assessment model and diagram offered later in the book specifically intended to enable social workers "to identify pertinent social systems" (p. 266)? What happened to the interacting physical environment? It literally disappeared from the picture and the model itself.

Lehmann and Coady (2001) defined a client's environment as "any aspect of the physical, social, and cultural environment, and what is most important will vary with individuals, time, and geography" (p. 72). This was positive. The physical environment was given as an important component of the overall environment for social work practice and a potential variable influencing human activity. For some unexplained reason, however, the accompanying diagram of this ecological perspective labelled social and cultural contexts while completely ignoring the physical environment.

Sheafor and Horejsi (2006) began their discussion of "social work's focus" with an assertion that the person-in-environment perspective "sets social work apart from other helping professions," and then defined the environment as "one's surroundings—that multitude of physical and social structures, forces, and processes that affect humans and all other life forms" (p. 9). I was impressed with such a broad notion of environment that included physical elements, natural settings, and the web of life surrounding human beings. A good start, but the authors then proceeded to make a distinction between the "immediate environment," defined in terms of social systems, and the "distant environment," which included features of "clean air, drinkable water, shelter, and good soil to produce food" (p. 9). For no apparent reason, nature had been relegated to the distance, but at least it was still part of the environment under consideration for social work. This promising beginning was rendered totally incomprehensible to me, however, when their illustrated model of practice featured a background labelled only as "the social environment" (p. 12), with no mention whatsoever of the physical environment. To make matters even more confusing, the accompanying assertion that "social work practice takes place within a *social environment* (sometimes called a *distant environment*)" (p. 14) completely contradicted their previous definition of "distant environment," which clearly included the natural world (air, water, soil).

Compton et al. (2005) defined the environment as "a combination of people and their social and environmental interaction in a particular personally, culturally, and socially constructed geographic space" (p. 48). Environmental interaction was distinguished from social interaction. Geographic space entered the equation, even if it was "socially constructed." How disappointing then to discover that the two diagrams intended to "illustrate the complexity of the interaction between person and environment" (pp. 49–50) did not include the physical environment. The authors directly acknowledged that their model "deals only with the social environment" (p. 48), but offered no explanation for dropping the previously important physical environment from the diagrams.

For Rothery (2002), the context of social work practice included relevant "aspects of the physical, social, and cultural environment" (p. 243). The ecological model as presented in the text clearly included "the myriad effects of the cultural, political, and physical context" (p. 244). Once again, the physical environment was affirmed as central in the written introduction to the practice model. How curious then that the accompanying illustration of people in context identified cultural and social factors, but failed to include the physical environment. In a later treatment of the ecological perspective, Rothery (2008) did include the physical environment in a diagram of "environmental elements to discuss in our effort to understand our clients' contexts" (pp. 109–110). This is a rare instance where the physical environment actually shows up in the pictorial representation of a social work model!

In general, it seems that we have some real problems holding to any focus on the physical environment in this profession. Like a strange uncle who makes everyone uncomfortable and somehow never gets into the family photograph, the physical environment seldom seems to make it into the diagrams of models of social work practice. Since such diagrammed practice models generally form the basis of the practice methods that follow, we might expect that the physical environment will also be absent from most assessment approaches in the social work literature.

ASSESSMENT AND THE ABSENT PHYSICAL ENVIRONMENT

Assessment is one of the foundations of generalist social work practice. A process of gathering and analyzing data, assessment precedes intervention at all levels of practice (Gold, 2002). Assessment involves "a search for full understanding of all significant facets of a client and the inter-influences that impinge, or appear to impinge, on the presenting situation" (Turner, 2005a, p. 17). If the profession has difficulty holding on to any fully developed notion of the physical environment in our theory base and models for practice, we might expect the role of the environment in conventional social work assessment approaches to be one of diminished importance.

Consider the formal person-in-environment (PIE) classification system. Much has been written about this social work assessment tool, which was created to parallel the *Diagnostic and Statistical Manual* for psychiatry and the International Classification of Diseases for medicine. Developed in the United States as an initiative of the National Association of Social Workers (NASW), the PIE system was published in the mid 1990s (Karls & Wandrei, 1994). A later development (Karls & Wandrei, 2000) allowed for PIE assessment reports to be generated using computer software, whereby the social worker enters codes and descriptors prescribed by the classification system. Four factors are used to describe a client's situation. Of particular interest to this discussion is "Factor II: Problems in the Environment." One of the originators (Karls, 2002) declared that "in PIE the environment includes both the physical and social contexts in which people live" (p. 196). This broad notion of environment is promising, but I find a familiar and limiting pattern as PIE assumes "social work's particular area of expertise as that of assessing and remedying problems in social relationships and problems in the social institutions of society" (Karls, 2002, p. 194). The organizing question for assessing problems in the environment is presented as "What problems exist in the social institutions in this community that are affecting the client?" (p. 195). While environmental factors such as food, shelter, and transportation are considered within Factor II, they are perceived only in terms of their economic or social implications.

Garvin and Seabury (1997) included the physical environment as part of the social context in which social problems and social work practice are found:

> The quality of the environment has only recently been emphasized as a focus of attention in social work as well as in many other professions. This is because we have reached a period in which the impact of human actions that negatively affect the environment cross state and national boundaries, and some affect the entire world.... These phenomena contribute to the problems people bring to social workers, and it is important for social workers who ignore their impact to stop doing so. (p. 15)

In spite of this plea for social workers to consider environmental phenomena, the assessment material presented in their book concentrated only on social functioning, social networks, and social roles.

Hoffman and Sallee (1994) affirmed the importance of social workers understanding "people as they interact with their environments" (p. 3), but their material on assessment directed workers only to "include data from the individual's social environment" (p. 190). Poulin (2005) offered elaborate instruments and worksheets for conducting a person-in-environment assessment, but nowhere did these assessment tools address the physical environment. Following a similar pattern, Sheafor and Horejsi (2008) initially identified "geography and environmental influences" as factors to be considered when developing a profile of the community during the intake phase of work (p. 236), but their whole chapter on data collection and assessment returned to an emphasis on "the various dimensions of a person's social function" (p. 288), with no acknowledgement of the physical environment beyond brief mention of home hazards such as "exposed electrical wiring" (p. 311) or a "gas leak" (p. 288).

The same flip-flop was evident in the material of Miley et al. (2004) on assessment. They began with an assertion that "social workers often locate resources in a client system's physical and social environments" (p. 243). Good, here was the physical environment identified as a source of resources for the client. The very next sentence, however, listed environmental resource systems as "family, friends, neighbours, co-workers, supervisors, classmates, teachers, social groups, school, church, social agencies, organizations, and the community." This list sounds like the social environment to me—no mention whatsoever of the physical environment, which has disappeared. No surprise then that the entire 96-page section on "Assessment and Planning" included only two paragraphs on "Assessing Physical Environments."

I was initially encouraged by Gilgun's (2005) recognition that "a person's environment is also physical" (p. 349). However, the tools presented for organizing assessment data were: the genogram ("helps organize both historical and contemporary data on the major figures in the client's interpersonal environment"); the eco-map ("a diagram of the family within its social context"); the stratification assessment ("the family's and the individual family members' place in the social structure"); and the developmental assessment ("stages of individual and family development") (pp. 351–354). All of these common tools for organizing data gathered for an assessment completely ignore the physical environment. There is little point in declaring the natural world to be an integral part of a person's environment if the assessment tools used are not capable of recognizing or incorporating these aspects.

The assessment format offered by Lundy (2004) made passing reference to the client's housing conditions and living arrangements. Workers were encouraged to consider questions such as "Are the living arrangements adequate, accessible, and affordable for all members of the family?" and "Do they live in a healthy environment?" (p. 120). Visual tools suggested for organizing the data gathered during assessment, however, were the standard genograms, eco-maps, social network maps, and life lines (pp. 120–124), none of which reflect information about the physical environment.

Johnson and Yanca (2007) drew on the work of Germain and Gitterman (1980) when they identified the "built world" and the "natural world" (p. 272) as components of the physical environment that are potentially open to manipulation by social workers. Their "Schema for the Study of a Geographic Community" (pp. 134–135) began with a general category of "Setting, History, Demography" that included such features as the location, size, relation to other geographic entities, transportation, settlement patterns, housing, mobility, population distribution, and physical location of businesses and institutions. Clearly the physical environment was considered an essential component for understanding community. A few pages later, the "Schema for the Study of a Social Agency" (p. 143) paid far less attention to the physical environment, mentioning only agency boundaries and financial value of the physical property. Finally, the "Schema for Development of a Social History: Individual" (p. 194) did not include the physical environment at all, beyond the client's street address as identifying information. All other environmental concerns were expressed in social terms. Why was the physical environment much less relevant for understanding the individual than it was for understanding the community? With such a limited assessment perspective, it came as no surprise that the "unit of attention" or focal point for social work intervention was declared to be "either a person, a social system, or the transaction between them" (p. 223). Once again, elements of the physical environment were defined as outside the scope of social work assessment and intervention.

I can almost feel the environment bubbling under the surface here, refusing to go away. It popped up again later in the Johnson and Yanca (2007) book when "environmental manipulation" was discussed as a legitimate strategy for social work intervention (pp. 271–275). Throughout most of the discussion the environment was expressed in social terms (formal and informal social systems surrounding the client), but there was also an intriguing consideration of spatial aspects of the environment (privacy and personal space, territoriality, proxemics, crowding, physical barriers to movement) and whether or not these physical realities were subject to change. Could they be appropriate targets or units of attention for social work? At least the question was raised.

Miley et al. (2004) suggested four questions for social workers to ask when assessing the client's physical environment (p. 258):

1.   In what ways does the client system's physical environment enhance or curtail its ability to function?

2.   How does the client system respond to the stress that results from factors in the physical environment?

3.   What modifications to the physical environment would be helpful to the client system? Which are most realistic to pursue?

4.   Who has the decision-making authority to ensure that these modifications will be made?

This is a positive step—an actual framework for consideration of the impact of the physical environment leading to possible environmental interventions. I do wonder, however, what happened to the notion of a transaction between the person and the environment. It seems to me that all of these questions could also be asked from the perspective of the physical environment in the foreground and human activity in the background:

1. In what ways does the client system's activity enhance or curtail the physical environment's ability to function?

2. How does the physical environment respond to the stress that results from factors of human activity?

3. What modifications to human activity would be helpful to the physical environment? Which are most realistic to pursue?

4. Who has the decision-making authority to ensure that these modifications will be made? What are the potential environmental and social consequences of not making these decisions?

CASE STUDIES: PHYSICAL ENVIRONMENT AS A RURAL CONSIDERATION

I have already shown how many popular textbooks have either ignored or minimized the physical environment in their assumptions and practice models. What about published case studies selected for social work students to apply their developing perspectives and skills? How have these case illustrations dealt with the physical environment?

Hancock and Millar (1993) published *Cases for Intervention Planning*, a source book of cases designed to help social work students learn to conduct appropriate assessments and develop intervention strategies. From a generalist perspective, they acknowledged that "it is conceivable that an integrated approach could be developed where a case is approached from a policy, HBSE [human behaviour in the social environment], and practice perspective simultaneously" (p. ix). Apparently, it was not conceivable for an integrated approach to include the physical environment. Influenced by the course structures in most schools of social work, the authors may have unconsciously directed social work students away from any consideration of the physical environment. Of the 30 case studies offered in the book, only two presented any details of the physical environment at all (Chapter 21 from the Beaver Creek Indian Reservation and Chapter 28 from a small rural community where transportation issues contributed to the problem). I perceive an indirect or unspoken message here that physical location and environment are only relevant aspects of a social work assessment and intervention with Aboriginal and rural peoples.

Rivas and Hull Jr. (2004) published the third edition of *Case Studies in Generalist Practice*, in which this pattern of rural and Aboriginal applications was confirmed but not explicitly identified. Three of the case studies included discussed the physical environment. Gottlieb and Gottlieb (2004) offered a rural case study, with much description of the rural Appalachian setting in a southern county of West Virginia. As crucial data for the assessment, the authors described roads, buildings, and water systems in the small rural town where the client family lived. Wahlberg (2004) presented a case study of work on a Native reservation, highlighting first impressions and reactions to the physical setting and then moving on to an appreciation of the local spirituality, "which stressed a holistic unification of people and the environment" (p. 69). Location and the construction of culturally appropriate structures were important components of the treatment program in this setting. Examining work with a native Hawaiian family, Mokuau and Iuli (2004) used the phrase "environmentally at risk" to categorize the family's living situation in a low-income housing project. Not only was the immediate environment a crucial part of the assessment, but the worker also had to be aware of larger issues involving cultural roots, identity, and land rights, all of which were connected with the specific geography and history of the region.

In their book *Connecting Policy to Practice in the Human Services*, Wharf and McKenzie (2004) included a case example from the work of the British Columbia Commission on Resources and the Environment (CORE). The case involved forestry and land use planning in rural regions of British Columbia. The authors openly acknowledged that "it may seem curious to feature an example concerned with environmental issues in a book focused on the human services" (p. 99), but they offered two justifications: (1) agreement on contentious land use plans might offer hope for the resolution of other difficult issues in rural regions; and (2) other disciplines might learn from the model of shared decision-making used by CORE in the case study. I note with interest that neither rationale argues for the interconnection of environmental issues and social policy. Here was an opportunity to make the case that land use planning cannot be separated from social policy, especially given the authors' definition of community as "a group of people having common interests and sharing a particular place" (p. 104). If community has a geographic location, then there should be no need to explain the relevance of a land use planning example in a human services book.

With the one exception of a single geriatric case study (Patterson, Jess, and LeCroy, 1999) appearing in *Case Studies in Social Work Practice* (LeCroy, 1999), where a senior's immediate physical environment (home, yard, and neighbourhood) was assessed as well as the social environment, I can find no mention of the physical environment in case study collections other than these rural and Aboriginal illustrations.

ENVIRONMENTALIST OR SOCIAL WORKER?

In the eighth edition of his popular textbook *Introduction to Social Work and Social Welfare: Empowering People*, Zastrow (2004) composed a list of 26 pressing issues to be addressed by social welfare in today's world (ranging from funding for services through to, amongst others, abortion, AIDS, affirmative action, and gun control). None of the identified concerns arose from environmental issues. In his broad introduction to social welfare, Zastrow did offer a single sentence addressing threats to civilization from "overpopulation, depletion of energy resources, excessive use of toxic chemicals, likelihood of mass famines and starvation, and dramatic declines in the quality of life" (p. 24), but these were presented as concerns of "environmentalists" not social workers. This is a most interesting distinction! Threats to the natural world and the associated quality of human life were seen to fall within the scope of environmentalism, but not social work.

I suppose this is the logical conclusion of perceiving the environment as a social environment. If social workers are concerned with only the social environment, environmentalists are left to tackle issues of the natural environment. *We* deal with social injustice, abuse, and oppression while *they* deal with environmental injustice, abuse and oppression. Of course, some social workers could also be environmentalists, but not necessarily. They might not be doing social work when they are tackling environmental issues. Environmentalism could be a sideline or hobby, activity outside the boundaries of social work.

In my opinion, Zastrow begins to reverse this position at the end of his lengthy book. The final chapter presents a comprehensive discussion of the population crisis and environmental problems as issues facing all humankind. Crucial matters such as population growth, food and water supplies, land degradation, energy, waste disposal, and pollution are examined with a sense of urgency. Confronting these problems, however, is again presented as the responsibility of "environmentalists" and "those concerned about preserving our environment" (p. 593) rather than social workers. But why include this material in a mainstream textbook at all if it is not relevant for social work? If these major environmental concerns fall outside the domain of social work, then why present them in such a compelling manner at the dramatic conclusion of a book on social work and social welfare? I think Zastrow is struggling with the very boundaries he defined for the profession, and struggling to include concerns for the environment as a legitimate focus for social work.

Zastrow (2004) ends his book with this appeal: "All of us can take steps to help save our earth. Whatever happens to the earth will surely affect everyone. This chapter (and this text) ends with challenging you to work for the improvement of human living conditions" (p. 599). This appears to be a strong call for individual social workers and students to take up environmental causes. I see this as a very important step in the mainstream social work literature. I can see that no clear case is made to include this urgent environmental work within the domain of the profession, but I wonder if that can be far behind if enough individual social workers do actually take up the environmental challenge posed by Zastrow.

Given the prevailing tendency in the mainstream social work literature to completely neglect the physical environment, this chapter has examined a few initial but largely unsuccessful attempts to acknowledge the importance of the natural and built worlds. Environmental rhetoric has been sabotaged by an inability to follow through with meaningful models and methods that integrate environmental considerations and concerns. Although relatively rare, there are instances of comprehensive attempts to place environmental issues at the core of social work theory and practice. The next chapter examines those efforts to fully incorporate the physical environment into social work.

CHAPTER THREE

CLEARING SPACE IN THE CONCEPTUAL BRAMBLES

Incorporating the Physical Environment

> When examining the relatively unexplored
> territory of ideas about the physical environment,
> interested social work explorers will need to pick their way
> through unmapped conceptual thickets and brambles
> to locate areas that can be cleared as practice sites.
> (GERMAIN, 1981, P. 103)

More than 25 years ago, Germain's poetic language offered an apt metaphor for the task of social workers trying to hold on to notions of the physical environment as central to practice. In Chapter 1, I examined how the mainstream social work literature has tended to dismiss all physical aspects of the environment, preferring to deal with only the social environment instead. In Chapter 2, I looked at some initial efforts that have run counter to this prevailing dismissal of the physical environment. Those authors voiced an intent to include the physical environment as central to social work, but their declarations were sabotaged by failure to include a developed sense of the physical environment in the practice models, diagrams, assessment tools, and case studies presented.

In this chapter, I look at the work of another group, relatively few in number, that has put forward a strong case for the importance of the physical environment in social work theory and practice. Returning to Germain's (1981) metaphor, these are the people who have worked to clear practice sites in the conceptual undergrowth—sites where the rest of us can pause and reflect on what social work might look like if we incorporate the physical environment rather than ignoring or minimizing its importance for our work.

Beginnings

Based on an ecological perspective borrowed from biology, *The Life Model of Social Work Practice* (Germain & Gitterman, 1980) clearly set out a framework for integrating social work methods and providing service based on client need. The ecological perspective provided a theoretical foundation as well as a useful framework for integrating levels and approaches to practice. According to these authors, the physical environment was a key component of the ecological perspective adopted by social work. As Gitterman (2002) later explained, ecology is "a biological science that examines the relation between living organisms and all the elements of the social and physical environments" (p. 105).

Germain (1981) emphasized the physical environment in an innovative chapter from the same period devoted entirely to "The Physical Environment and Social Work Practice." Observing how the profession was adapting the ecological perspective, she was alarmed that the physical environment was

> still a largely unexplored territory in social work practice and tends to
> be regarded—when it is regarded at all—as a static setting in which
> human events and processes occur almost, if not entirely, independently
> of the qualities of their physical setting. (p. 104)

For the mainstream profession, the physical environment had become static, a lifeless backdrop for human activity, much like a simple stage setting for scenes from a play. Germain (1981) strongly opposed this development, arguing that the physical environment was complex and multi-textured. She presented the physical environment in terms of the natural world (animate and inanimate nature) and the built world (structures and objects constructed by humans); these environments were further textured by time (daily, seasonal, and annual rhythms; social cycles of time created by human activity) and space (open and closed) (p. 105).

Germain was not, however, swinging to the other end of the pendulum and declaring primacy of the physical environment over the social environment. Rather, she asserted the interdependence of the physical and social environments. People actively experienced the physical environment; it held subjective meaning for them. People perceived their environment according to their own experiences, individual and cultural. "People's perceived (subjective) physical environment may differ from the actual (objective) environment, or from the subjective environments of others who share the same physical setting" (Germain, 1981, p. 109). Germain did, however, acknowledge the potential utility of theoretical separation of the physical and social environments for purposes of assessment and analysis, but she warned of a danger of "oversimplification" (p. 105). I suspect she was reacting to the trend in the larger profession towards simply dismissing the physical environment once it had been defined as separate.

Writing at the same time in NASW's journal *Social Work*, Weick (1981) also decried social work's focus on human behaviour to the neglect of the physical environment. He thought this neglect of the physical environment might be explained by the forces of "industrialization, science, and psychoanalysis" (p. 140). Possibly the post-war trend towards technology and urbanization diverted attention from the natural world and our interdependence. Control of the environment through technology was the new focus. The growing fascination with personality and the inner workings of the mind may have also served to relegate the physical environment to the background.

Similar to the work of contemporaries Germain and Gitterman, Weick (1981) argued that the environment component of the person-in-environment paradigm must be viewed as multidimensional. Weick included both internal and external factors in a matrix of four possible environments: internal social (conventional social work micro practice); external social (conventional social work macro practice); internal physical ("genetic traits, metabolism, organ functioning, and adaptive capacity"); and external physical ("climate, air, noise, food, biological rhythms, and atmospheric conditions") (p. 142). Together, these environments "form a dynamic matrix of interaction that shapes an individual's behaviour.... To view behaviour accurately, all four sets of environmental influences must be considered" (p. 142). Here was a simple yet powerful matrix for assessment that included both social and physical environments. This was in our literature some 25 years ago, but it was not picked up or developed by the mainstream textbook authors who, for reasons discussed in Chapters 1 and 2, continued to ignore the physical environment in their models for assessment and intervention.

With the possible exception of Soine's (1987) appeal to include content on poisons and hazardous materials in the curricula of schools of social work, it seems the call for our profession to take the physical environment seriously languished for more than a decade in the literature until the debate was renewed in the mid 1990s.

MID-1990s FORUM

During the 1990s, the physical environment was generally ignored or mentioned only in passing in mainstream social work textbooks. Several examples of this pattern have been given as evidence in Chapters 1 and 2. At the same time, however, an important discussion on the subject appeared in the NASW journal *Social Work*, which published four articles between

1992 and 1995 that made a strong case for inclusion of the physical environment within the domain of social work.

Describing the social work literature devoted to the physical environment as "sparse" (p. 391), Gutheil (1992) was concerned that this oversight could result in social workers' neglect of physical surroundings when conducting assessments. "Because social workers do not attend to clients' physical environment with the same precision they bring to the social environment, they underuse an important tool in assessment and practice" (p. 395). The article focused on the immediate physical environment of the client (lighting, furniture placement, fixed space arrangements within buildings), and concentrated on the influence of internal design on interactions and relationships. Several concepts were put forward to inform social work practice, such as territoriality, personal space, sociopetality (welcoming environments), and sociofugality (stressful environments that discourage relationships). Mostly these were borrowed from sociology and environmental psychology. The overall effect of these potential additions would be for social workers to broaden their scope of practice to include assisting clients to negotiate obstacles and challenges to control of their immediate physical space. Questioning the choices made by those who designed physical spaces would open up a whole new area of interdisciplinary work, potentially bringing social workers into collaboration with designers and planners.

The next year, Hoff and Polack (1993) considered human–environment interaction from the other side. Rather than looking at how the environment influences human activity, they emphasized "the human threat to environmental viability" (p. 208), which they claimed had been ignored in the social work literature. They argued that "a sustainable relationship with our physical environment is the necessary foundation for continued economic well-being" (p. 209). If social work could expand its ecological model to incorporate a much broader concept of environment then the degradation of the natural world might become a legitimate focus for our profession, which "has the theoretical base and practice skills to respond to the social dimensions of environmental issues at the local, national, and international levels" (p. 209). Here was a powerful new idea. Based on our theory and practice experience from a person-in-environment perspective (even if we had limited this to the social environment), we might have something to offer as the world was beginning to notice environmental degradation and resource limitations.

Later that year, Berger and Kelly (1993) also called for social work's ecological model to be extended "to a full awareness of humans' role in biological as well as social ecosystems" (p. 524). Arguing that the foundation values of the profession would also need to be expanded to support this new direction, they developed a 12-point "ecological credo for social workers" (pp. 524–525):

1. Social work is concerned not only with the interactions between people and their social environments, but with the full range of interconnectedness among all systems within Earth's biosphere.

2. Social work promotes self-determination and respect for individuals within the context of individual and community respect for nature. Self-respect and respect for nature are inseparable.

3. Social work believes in global equality, that is, in the right of all people of the world to share equally in Earth's bounty. It recognizes that global harmony cannot exist when a minority of people in developed nations consume a disproportionate share of global resources.

4. Social work seeks the establishment of social and economic policies that promote human welfare. Human welfare is understood to include not only short-term needs for consumption but also the needs for future generations. Therefore, social work sup-

ports only those social and economic policies that promote sustainable use of Earth's resources.

5. Social work has the responsibility to promote social, political, and economic systems that respect the integrity of the biosphere. This support extends to new means of economic, social, and political organization that will reverse current practices of ecological damage and resource depletion.

6. Social work is confident of the integrity of the natural ecosystem. At the same time, social work acknowledges the carrying capacity of the biosphere and respects the limits of that capacity.

7. Social work values the principle of diversity. The diversity of ecological niches and life forms that form the biosphere is reflected in the diverse races, ethnic groups, cultures, and values of people. Such diversity is valued for the resilience it brings to all systems.

8. Social work assumes a global and universal perspective. Humans are not separate from, nor superior to, other parts of the biosphere. Rather, humans are but one aspect of a vast universe in which every aspect is interconnected.

9. Social work promotes stewardship of the Earth's resources by its human inhabitants.

10. Social work acknowledges the obligation of its professionals to speak out when they have knowledge of damage to the environment that will adversely affect the quality or sustainability of life for current or future generations of living systems.

11. Social work believes that humans have the moral capacity to apply their intelligence and technology to create ecologically sound, humane, and sustainable lifestyles.

12. Social work believes in the essential goodness of people. The people of Earth will voluntarily live in harmony with Earth's resources when afforded the opportunity to assume ecologically responsible lifestyles.

Two years later, Berger (1995) voiced the appeal once again in a provocative editorial that applied the label "habitat destruction syndrome" to a global illness whereby "the human race is collectively engaged in practices that damage the environment and ensure our eventual self-destruction" (p. 441). The persuasive argument was that we have become desensitized to the threats to our environment and immobilized by a fear that the problem is too big for us to handle. Assuming that habitat destruction needs to be understood as the "greatest threat to our social welfare," Berger again asked why we do not "add environmental activism to social work's list of social welfare concerns" (p. 443). Presenting evidence of the degradation of water, air, soil, and atmosphere (with the resulting negative impacts on human health and economic welfare), Berger suggested that there are many things social workers could do, including (pp. 442–443):

- Learn about environmental issues (immediate environment and global concerns);
- Learn about environmental problems and oppression (toxic dumps in poor neighbourhoods or Third World settings);
- Speak out against environmental injustice;
- Make and promote lifestyle changes (conservation, recycling, public transit, environmentally responsible investments);
- Engage in collective political action.

Taken together, the ecological credo (Berger & Kelly, 1993) and the suggested actions for social workers (Berger, 1995) presented social work with a challenging new path. New language and concepts had been introduced, with references to respect for nature, natural diversity, sustainable use of resources, environmental oppression and injustice, ecologically responsible lifestyles, and obligation to and stewardship of the Earth. This new language brought forward an eloquent and powerful challenge to the profession of social work, rooted in a new understanding of our partnership with nature. Yet I have found little evidence that the mainstream profession and literature even considered the ecological credo and its revolutionary stance. Perhaps it was too much too soon. Possibly the profession, in an era of neo-conservative policy and service cutbacks, could not afford to consider expanding its scope.

ENVIRONMENT AS CENTRAL

In 1994, Hoff and McNutt published a book called *The Global Environmental Crisis: Implications for Social Welfare and Social Work*. Motivated by the environmental threats facing humankind, the authors began with the premise that human and environmental welfare are "inextricably linked" (p. 2). Hoff and McNutt (1994) argued that social work and other professions will have to move beyond outdated goals of individual well-being and social welfare to adopt new models geared more towards sustainability and protection of the environment. Contributions to the edited volume made a strong case for the integration of social justice and environmental justice. The case studies explored the impact of environmental destruction on oppressed populations in various regions of the world. For example, Lovell and Johnson (1994) offered specific adaptations to practice at the micro level for working with victims of environmental hazards and with the often-neglected affluent consumer whose lifestyle must change. Macro-level concerns were discussed in terms of environmental citizenship (Kauffman, 1994) and environmental social action (Shubert, 1994).

Challenges to social work education were also addressed, with specific attention to field practicum placements that aim to mitigate environmental hazards (Rogge, 1994) and expansion of "human behaviour and the social environment" curricula to include the physical environment (Kauffman, Walter, Nissly, & Walker, 1994). With its declared emphasis on "relationships between people and the natural environment at all systemic levels" (p. 10), the Hoff and McNutt (1994) book was ahead of its time. The passage of time has not lessened the importance of their message.

In a collection of readings from Australia, McKinnon (2001) contributed a chapter entitled "Social Work and the Environment" in which she argued that social workers cannot continue to ignore the natural world because environmental degradation is now perceived as a social justice issue. The social environment becomes irrelevant at the point where a polluted and degraded physical environment is unable to sustain life. "Social work as a profession can no longer ignore the intertwined nature of social, environmental, and economic systems" (p. 193). She called for an integration of social work and environmental goals leading to an overall "earth-care awareness" (p. 203) and ultimately "ecologically sustainable development" (p. 205).

In a later version of the chapter, now called "Social Work, Sustainability, and the Environment," McKinnon (2005) observed the growing public awareness of green issues through environmental associations, organizations, and political parties. However, she also noticed that social work is generally on the outside of this new discourse on the environment and sustainability. Her review of major social work journals in the decade following release of the report of the United Nations World Commission on Environment and Development (1987), often referred to as the Brundtland Report, uncovered only seven articles with reference to the environment or sustainability in their titles. While environmental issues have "slowly but surely infiltrated the everyday thinking of the social mainstream," sadly "there has been little social work interest in the links between our professional focus on social systems and the broader economic and environmental systems" (p. 226).

Arguing that the profession of social work needs to move beyond social relationships to consider the entire life context as the focus of practice, some authors have advocated an "eco-social" approach (Matthies, Jarvela, & Ward, 2000; Narhi, 2004; Tester, 1994). As one of the foundation principles for eco-social work, Payne (2005) identified "promoting positive use of natural resources and self-consciousness about life styles respectful of environmental resources" (p. 155). Such awareness of the environmental implications of lifestyle choices on a personal and societal level served as the foundation for an important book by Coates (2003), who called for nothing less than social transformation and a new ecological paradigm for social work.

In his book *Ecology and Social Work: Toward a New Paradigm*, Coates (2003) argued that the Western focus on the individual and competition has made us blind and indifferent to our connectedness with the natural world. Through our commitment to continuous economic growth, we have created an ecological crisis "where human activities exceed nature's capacity to replenish and regenerate" (p. 20). Facing this crisis will require a "new set of values and beliefs, a new world view" (p. 37) to establish healthy connections with our sustaining natural world.

Surveying social work's historical lack of attention to concerns of the physical environment, Coates labelled modern social work as "domesticated" and "co-dependent" (p. 38), a partner with economic development that helped people fit in with unrestrained development processes rather than to challenge them. Coates advocated a new role for social work in helping people to recognize the ecological crisis, question the underlying values of consumption, and then prepare themselves for "individual and collective global consciousness [that] needs to be integrated with the innate wisdom of the Earth" (p. 77). His paradigm calls for social work to become a major player in the transformation of society towards global consciousness and environmental well-being, not simply economic or social well-being. The alternative would be to continue contributing to the destruction of the natural world. Coates presented the alternatives in this way: "Social work has the choice of continuing to support a self-defeating social order or recreating itself to work toward a just and sustainable society" (p. 159).

In order for social workers to prepare for their role in the proposed transformation, Coates suggested they will require "a global consciousness" (p. 151) as well as broader understanding of relationships between specific human activities and the Earth. The first step in this new approach to social work education and training involves reviewing our profession's role in the current system and the patterns and ideologies that inform our models of practice.

Articulating the philosophical and value base for Canadian social welfare, Delaney (2005) asserted that "human potential for development responds to and affects the physical and social environments" (p. 18). We have seen this type of declaration before in the social work literature, only to have the physical environment quickly disappear through smoke and mirrors. That did not happen this time. Delaney (2005) maintained the balance. Understanding our interactions with the social environment was based on a notion of mutuality; understanding our reciprocity with the physical environment was based on commonality—"the physical environment is the common place to sustain all of the world's life forms" (p. 18). This time the physical environment even makes it all the way into the diagram, where commonality, mutuality, and equality are presented in a visual arrangement as societal values that support social welfare (p. 21).

In the same book in which Delaney illuminated the philosophical and value base for Canadian social welfare, I attempted to make the case for a geographic base (Zapf, 2005b). Social welfare had been presented at the outset of the book as pursuing a vision of individuals achieving "their full human potential and to do so in a manner they can accept while showing due regard for the rights of others" (Turner, 2005b, p. 2). I argued that such due regard for the rights of others

> may have to be expanded to include a regard for the rights of the environment, an attitude of respect and stewardship rather than exploitation, an understanding that "full human potential" must exist beyond individual achievement. Full potentials for individuals, societies, and environments are not separate goals; they are expressions of the same desirable healthy state. (Zapf, 2005b, p. 70)

Perhaps Compton et al. (2005) expressed it most directly with their simple observation:

> We cannot view the environment as something separate from and distinctly outside ourselves. People do not occasionally bump into the environment. Rather, the environment is part of us and we are part of it. Air and water are inside as well as outside our bodies, and we depend upon the environment for our very survival. From the time of birth, the environment affords us the material from which we construct our lives. The world provides the context in which we make decisions and undertake our journeys. (p. 52)

A recent social policy text stands out in terms of its recognition of the physical environment. Graham, Swift, and Delaney (2009) placed the "environmental imperative" clearly at the top of the list of essential issues that will shape social policy in the 21st century. They hold the developed world fully responsible for the present ecological crisis. Our "*ecological footprint* represents the impact of each person on the environment, through consumption of fossil fuels, agricultural production of food and other products, misuses of forests and water, human settlement, and so on" with results that are "potentially calamitous" (p. 100). Acknowledging small-scale immediate actions such as recycling and taking public transit, Graham et al. (2009) warned that "social policies have scarcely begun to integrate issues of social justice for people with the imperative of physical ecology" (p. 102). They suggested that the profession of social work missed the boat by applying ecological theory only to issues of people and their social contexts, as a "method of integrating direct practice with social (or as social work uses the term, 'environmental') issues" (p. 154). What may be necessary to move past this block is "a new world view that equates spiritual growth and self-realization with stewardship of the earth" (p. 102).

This new world view has been affirmed in a strongly worded policy statement from the American NASW (2000):

> Protecting people and the natural environment through sustainable development is arguably the fullest realization of the person-in-environment perspective. The compatibility of sustainable development and the person-in-environment perspective is a firm theoretical foundation from which to apply macrolevel social work practice to person–natural environment problems. (p. 105)

This position was supported by an entry on "Environmental Issues" appearing in the *Encyclopedia of Canadian Social Work* (Turner, 2005c) that acknowledged the interconnection of environmental, social, economic, and political concerns:

> Weather change, depletion of natural resources, pollution and environmental ecosystem conflicts have challenged the quality of life for people all over the world ... depletion of such natural resources as forest, marine life, and fresh water is beginning to harm some Canadians. Environmental depletion and change are closely associated with conflict as well as poverty. Reduction of supplies such as forest and fish stocks have created harm as well as policy disputes. Increased environmental disasters such as floods and severe weather, owing in part to global warming, have created trauma as well as financial losses. (Bohm, 2005, pp. 122–123)

The centrality of the physical environment to macro-level practice has been given a boost in a recent textbook entitled *Human Behavior and the Social Environment: Macro Level—Groups, Communities, and Organizations* (van Wormer, Besthorn, & Keefe, 2007). I wonder why the book's title was phrased in purely social terms, giving no hint of the importance of the physical environment for human activity, when the content includes an entire chapter devoted to "Human Behavior and the Natural Environment: The Community of the Earth" (pp. 222–262). Here, in a mainstream social work textbook, are discussions of biodiversity, global warming, war, and consumerism presented as challenges to the planet and our profession. Our history of ignoring the natural environment is exposed and alternative perspectives are presented, resulting in an "expanded ecological social work model" that "presents social work with the opportunity to take a philosophically grounded position that publicly and openly acknowledges an awareness of the interrelatedness of social, political, economic, and environmental issues" (p. 255).

PLACE AND SPACE IN SOCIAL WORK

As social workers, we are often encouraged to think globally and act locally, to be aware of broad issues of injustice and related oppressions while challenging their manifestations in the immediate settings in which clients live their lives and we do our work. In this chapter, I have attempted to share examples from the social work literature in which authors have called for the physical environment to be included at the core of our profession's approach to fighting oppression and injustice. Much of this work has been conceptual and grand, considering global environmental threats and the Earth in general. I have found some evidence, although far more rare, of efforts to bring these concerns to the local level, to locate them geographically. In this material, I find the beginnings of a sense of place in the social work literature.

In their discussion of key concepts in the ecosystems approach, Allen-Meares and Lane (1987) argued for the inclusion of the physical environment as a variable in their assessment model. Others have also done this, as we have seen. What sets this work apart, however, is a declaration that human behaviour must be recognized as "site specific" (p. 518). How I act or react in one setting may be very different from how I act or react in another setting. At first, this may appear to be simplistic and obvious, but the implications are profound. Behaviour can be located in the physical environment. Understanding behaviour requires an understanding of the physical context, not just the social context. The question becomes: What does it mean to act this way in this place?

Chaskin (1997) considered this same notion in a review of the literature on community and neighbourhood. Physical location was found to be optional for "community," which was based on "connections: some combination of shared beliefs, circumstances, priorities, relationships, or concerns ... [which] may or may not be rooted in place" (p. 522). As we are seeing, especially with the internet, communities of interest do not have to share a physical space to maintain their group identity and interaction (although the spatial component of cyber*space* offers intriguing possibilities!). On the other hand, "neighbourhood" was discovered by Chaskin (1997) to be very much "a spatial construction denoting a geographical unit in which residents share proximity and the circumstances that come with it" (pp. 522–523). This work suggests that a sense of place or geographic circumstances—including the resources and challenges of the proximate environment—is a defining feature of neighbourhood and a key factor for understanding human behaviour. After all, "the person you know on the Internet won't feed your cat or watch for suspicious characters while you are away on vacation" (Locke, Garrison, & Winship, 1998, p. 283).

A new direction in community development, usually labelled asset-based or capacity-focused development (Kretzmann & McKnight, 1993), has paid attention to the assets or capabilities of a community, rather than its needs, deficits, or problems. Recognizing and mobilizing these local capacities would be the key to community building or rebuilding. This approach still placed emphasis on social aspects of community. "Individuals, associations and institutions—these three major categories contain within them much of the asset base of every community" (Kretzmann & McKnight, 1993, p. 8). However, the physical environment was still considered with a call to "highlight other aspects of a community's assets including its physical characteristics—the land, buildings, and infrastructure upon which the community rests" (p. 8). Buildings and recreational land even made it into the diagram of a "community assets' map" (p. 7) and the visual representation of physical resources (p. 313). The latter considered vacant lots, vacant buildings, vacant commercial/industrial space, and underutilized space as inputs, with community playgrounds, neighbourhood cultural centres, community gardens, single-family housing, multi-family housing, business incubator centres, ethnic museums, and theatres as potential outputs.

> Abandoned and vacant properties have always been thought of as liabilities: they are unsafe, they absorb precious resources and they inhibit economic and residential development. What would happen if communities began to perceive those same properties as potential assets? A vacant lot filled with junk does have the potential of becoming a community garden. A vacant school becomes a combined-use living and learning centre. An abandoned industrial site becomes a small business incubator. And an abandoned warehouse becomes a theatre, complete with retail, rehearsal and workshop space. The possibilities go on and on. (Kretzmann & McKnight, 1993, p. 312)

Building on the work of Kretzmann and McKnight (1993), Diers (2004) reported on several community-development projects initiated through the Department of Neighborhoods in Seattle. He made a clear distinction between a community and a neighbourhood:

A neighborhood is not the same as a community. A neighborhood is a geographic area that people share, while a community is a group of people who identify with and support one another. It is possible for a neighborhood to lack a strong sense of community, and conversely, it is possible for there to be a strong sense of community among people who don't share a neighborhood. A community can be defined by a common culture, language, or sexual orientation, regardless of geography. (p. 170)

It would appear from this distinction that communities need not have a geographic base, basis in place, or defined physical space for interaction. I find it curious then that on the next page, Diers observed the following:

Community initiatives generally have a positive effect on the environment. While academicians struggle to define and measure *sustainability*, strong communities tend to practice sustainability whether or not they have ever heard of the term. In communities, people care for one another and the place they share. (p. 171)

Through this observation, community has been tied very closely with geography, with place. Maybe community can be placeless in the abstract, but caring and supportive interactions occur in context, most often a shared place, literally a common ground (Zapf, 2008).

Returning to the language of Germain (1981), I see these rare attempts to acknowledge a sense of place in the social work literature as small sites that have been cleared in the conceptual brambles. They are few, fragile, and vulnerable to being overgrown or trampled. There is also a danger within the Western mindset that place can quickly become perceived as property, a commodity. Bishop (2002) effectively identified this process when she described an exercise she uses with groups to explore issues of power and oppression:

In order to introduce the concept of private property, it will be essential to introduce a belief system that portrays the earth, animals, plants, and resources as "things," separate from large ecological systems, with no value except what they can be used for by their owners. (p. 26)

Tracing the historical foundations of oppressive processes, Bishop (2002) looked to the 16th century:

Western culture was undergoing a profound ideological revolution. The idea was taking hold that the earth is a dead piece of rock, or a machine, valuable only as raw material for human use. From this basic notion, new understandings emerged. Land was beginning to be thought of as a commodity, to be bought and sold.... The landowners began to pressure for enclosure and fencing, so that they could use the land for mining or raising sheep, cattle, and crops for sale. They worked to transform the land from a resource with multiple uses for the whole community to private property for private profit. (pp. 38–39)

Far from being an historical artifact, this private property mindset is active today:

Our transportation infrastructure—including even the sea bed under our harbours—our formerly public parks, public meeting places, prisons, public services, the seeds we grow, even our human genetic code are all being declared private property and sold to the highest bidder. (p. 47)

In a comprehensive discussion of social exclusion, Gingrich (2003) also looked at the political and power issues that may be actively associated with a sense of place. Physical location was a necessary component in processes of "geographical or spatial social exclusion" (p. 15). Geographical dislocation and segregation of identifiable groups of people had served historical purposes of exclusion, marginalization, and sometimes genocide. Gingrich (2003) offered examples of cultural diaspora and the creation of reserves for Aboriginal peoples. Social exclusion by geographic space is also evident in the spatial distribution (or polarization) of income zones in modern urban regions.

Overall, what can be said about our first attempts to understand place in the social work literature? We find a dynamic sense of context, of behaviour that is situated in places. Proximity, common ground, and shared aspirations contribute to neighbourhoods where physical features may be seen as assets or resources. At the same time, there is the danger that places can be perceived and owned as economic commodities, or used for purposes of oppression and social exclusion. We have little experience with the concept of place in mainstream social work and we are only beginning to appreciate its potential.

In this chapter, we have looked at a few social work authors from the 1980s who attempted to incorporate the physical environment at the core of their practice models, counter to the prevailing pattern of the mainstream profession. A lively debate in the mid 1990s connected this work with a growing awareness of an environmental crisis threatening human health and survival. Coates' (2003) book challenged us with a vision and the beginnings of a language for the real integration of social justice goals with concerns for a healthy and sustainable natural world. A sense of place has been suggested as potentially important for social work practice in specific contexts. While these notions of stewardship, sustainability, and place may be new to the mainstream profession, they have some standing at the margins of social work. The next three chapters explore understandings of the physical environment within the specializations of rural/remote social work, spirituality and Aboriginal social work, and international social work.

CHAPTER FOUR

RURAL/REMOTE SOCIAL WORK

Environmental Context and Place

> Geography is not destiny. However, where a
> person or family lives does have a bearing on what
> kinds of opportunities and services are available and
> on how isolated the person and community
> perceive themselves to be.
>
> (LOCKE ET AL., 1998, P. 72)

n an earlier chapter, I observed that most of the case studies in the mainstream literature that incorporated the physical environment into their assessments and intervention strategies were rural in focus. These rural case studies considered issues such as land rights, local water and transportation systems, land-based cultural identity, and land use planning. Rural concerns have generally been relegated to the margins of mainstream social work, where they have experienced an uneven cyclical history of attention and neglect. Periods of attention have resulted in the evolution of rural social work as a specialization within the larger profession. Arguing that isolated wilderness areas go far beyond conventional notions of rurality, some writers have made the case for a further specialization of remote practice.

Rural/remote social work voices are still within the profession, but they are at the margins or fringe. Just what is rural/remote social work? How is the physical environment perceived and approached by social workers outside of urban regions? Are there any lessons here that might help the mainstream profession to incorporate environmental issues in its mission? Attempting to answer these questions will be the focus of this chapter.

RURALITY AND RURAL SOCIAL WORK

In the North American social work literature, rural social work was recognized back in the 1930s. Much of the population was rural at that time. Given the economic depression and the dust-bowl conditions of that era, it is no surprise that provision of resources and services in rural regions was high on the agenda of the young social work profession. In 1933, Josephine Brown published *The Rural Community in Social Casework*, the first book on social work practice in rural areas. The journal *Rural Society* published numerous articles on rural social welfare throughout the 1930s. Then a strange thing happened. Rural social work literally disappeared from the literature for three decades. I have attributed this disappearance of rural social work for 30 years to a combination of factors: the addition of analytic psychology to our knowledge base and the subsequent fascination with clinical practice; the concentration of social work training in urban graduate schools; and a general post-war trend towards urbanization of the population, culture, and economy (Zapf, 2002).

Whatever the reason, the fact remains that I do not find any significant reference to rural practice in the social work literature from the 1940s, 1950s, and 1960s. It appears that the young profession through that period was seeking coherence and unity through consolidation of common values and practice foundations. The clinical focus on urban practice seems to have been the dominant or exclusive interest until 1969, when Leon Ginsberg addressed the American Council on Social Work Education (CSWE) to condemn the rampant neglect of rural communities as a subject for education and research in social work (Ginsberg, 1969). His speech opened the door for a decade of activity in the 1970s as a strong case was made for distinguishing rural social work from the obvious urban emphasis of the mainstream profession. This decade is often viewed as a golden era for rural social work in North America, featuring such noteworthy achievements as formation of both the American Rural Social Work Caucus and the Canadian Rural Social Work Forum, appearance of the first entry on rural practice in the *Encyclopedia of Social Work* (Ginsberg, 1971), publication of first edition of Ginsberg's (1976) collection of readings *Social Work in Rural Communities*, establishment of a CSWE Task Force on Rural Practice in the United States, initiation of an annual Institute on Social Work and Human Services in Rural Areas conference, and publication of the first issues of the journal *Human Services in the Rural Environment*.

The new rural social work knowledge base included many efforts to explore rurality and define rural areas. No longer satisfied with a simplistic urban/rural distinction, many writers began to divide the rural practice context into further subcategories (e.g., farmland, rural non-farm, villages, hamlets, boomtowns, metro-transitional communities, non-metropolitan regions, wilderness, recreation areas), with arbitrary population thresholds for defining rural areas ranging from 1,500 to 50,000 people (Zapf, 2001). Through initial research efforts, anecdotal practice accounts, and some theorizing, rural social work attempted to define itself as a valid specialization that could be clearly distinguished from mainstream urban practice. By the end of the 1970s, this goal had been accomplished. In a 1981 overview of rural social work as a movement, Martinez-Brawley concluded that "rural social work had succeeded in gaining a place in the ranks of the profession" (p. 201). O'Neill and Horner (1984) later referred to rural social work as a "contextual specialty" (p. 2) requiring special knowledge of the rural context, role definition appropriate for rural practice, and cautious application of conventional social work knowledge and skills.

Through the 1980s and 1990s, rural social work continued to develop as a specialization through ongoing conferences with published proceedings, journal and book publications, and the development of rural social work courses at educational institutions. Johnson (1980) published a collection of rural practice readings entitled *Rural Human Services*. Two years later, a textbook entitled *Rural Social Work Practice* appeared (Farley, Griffiths, Skidmore, & Thackeray, 1982). Martinez-Brawley (1990) later contributed *Perspectives on the Small Community: Humanistic Views for Practitioners*. Two further editions of Ginsberg's (1994, 1998) *Social Work in Rural Communities* were published. In Canada, the first edition of Collier's book *Social Work with Rural Peoples* appeared in 1984, with a second edition in 1993. From Australia came Cheers' (1998) *Welfare Bushed: Social Care in Rural Australia*.

In much of this literature base, particularly the American sources, the specialization of rural social work assumed a goal of modification or adaptation of urban programs and services for populations outside of urban areas. Cheers and Taylor (2005) labelled such an approach as the "mainstreaming" of rural needs, which involved "viewing them as identical to urban needs and responding with the same kinds of services and resources" (p. 243). Overall, rural social workers were expected to encounter geographic barriers and community resistance as obstacles to be overcome in providing rural peoples with access to the same level of service available in the city. These rural social workers were generally depicted as operating under conditions of high visibility, employed by public multipurpose agencies to provide informal and personalized services with few professional supports. Generalist practice has always been the preferred practice approach and, in many places, the only realistic option for rural social work (Bodor, Green, Lonne, & Zapf, 2004; Collier, 2006; Ginsberg, 1998; Lohman, 2005; Zapf, 2002).

Although the developing rural social work specialization was primarily focused on overcoming deficits experienced by rural populations, there were calls for new perspectives to integrate the welfare of the rural physical environment with the well-being of its inhabitants. The 1994 International Conference on Issues Affecting Rural Communities passed a Statement of Principle that:

> A wide variety of views and new ways of thinking on how best to ensure the economic, social, cultural, educational, environmental and physical well-being of rural communities be encouraged and supported in commissioned reports, in academic papers, in conferences, and in gatherings of citizens. (McSwan & McShane, 1994, p. 437)

DIVERGENT PATHS FOR RURAL SOCIAL WORK

In the last decade, there is evidence that things are changing again within the specialization of rural social work. At least three divergent paths or directions have emerged. One group has looked to advances in technology and transportation and concluded that maybe rural areas are not so isolated anymore. Rural peoples may have a local sense of focus, but they are subjected to the same international economic forces and social pressures as the rest of society. In a controversial article suggesting that rural social work has become an anachronism, Mermelstein and Sundet (1998) posed the question this way:

> Are things so basically different about the urban and rural contexts that we can justify a specialty, a title, a journal, a special alcove in the halls of social work? Increasingly, the evidence suggests "NO."
>
> Much of the defense for this specialization is rooted in the concept of the rural-urban continuum contained in the classical sociological studies and governmental panel reports on rural communities and/or traditional societies.... Those studies and reports have histor-

ical significance and, perhaps, are useful guides to some third-world countries but they have little relevance to the reality of 21st century America. (p. 64)

In contrast to this suggestion of the possible irrelevance of a rural context specialization, a second group has attempted to develop a new emphasis within rural social work. Scales and Streeter (2004b) assembled a book of readings entitled *Rural Social Work: Building and Sustaining Community Assets* in which they tried to incorporate the macro approach of capacity- or asset-building (Kretzmann & McKnight, 1993) within rural community work. Essentially, their approach was a reaction against the long-standing tradition of defining rural areas in deficit terms (what is lacking compared with the city). Instead, they focused their attention on rural community assets, including "the skills of local residents, the power of local associations, the resources of public, private, and non-profit institutions, and the physical and economic resources of local places" (Scales & Streeter, 2004a, p. 2). Rural social workers were to "consider the social problems of rural populations as stemming from both a physical environment and from a sociocultural or rural lifestyle perspective" (Daley & Avant, 2004, p. 37). The physical environment was not only a consideration for understanding rural problems, but local human and physical resources were also essential elements for the resolution of problems through community capacity-building (Ersing & Otis, 2004).

Are these two the only possible options for rural social work now? Must we either abandon rural social work as irrelevant or transform it into rural community capacity-building? I see another direction, although it is not easily identified as rural social work. A third group has attempted to take the rural social work focus on context and the physical environment and interject this into mainstream practice models. As a successful example of this process, consider a social work textbook by Locke et al. (1998) entitled *Generalist Social Work Practice: Context, Story, and Partnerships*. From its title the book appears to be a conventional generalist practice text, yet all three authors have extensive backgrounds in rural social work theory, practice, and education. While I do not want to suggest anything secretive or conspiratorial here (I am not trying to "unmask" the authors as secret ruralites!), I do think a plausible argument can be made that their book is a reverse smoke and mirrors operation. This time, context and even the physical environment have been snuck unannounced into the mainstream literature!

Central to the model of Locke et al. (1998) was the notion of context: "where social work gets done—the specific nature, qualities, and characteristics of a locality, which interact to shape the experiences of persons and groups and their sense of what is possible" (p. 3). Examining the social work principle of "starting where the client is," the authors observed that this directive usually means assessing an individual's motivation for change, resources, abilities, and pressing needs—all internal factors. They wanted to expand this focus to include the external "dimensions and forces of the context" (p. 15), including the physical setting. Concerned that the word *community* has over time "come to lose its association with place," the authors preferred to use the word *locality* "to help us define the physical nature of context.... Defining the physical location as locality allows us to better understand the complexities of place" (pp. 68–69). Interactions between geography and culture help to explain the meaning that places have for people and the identity perceived by the residents of a locality. "Geography shapes life experiences, defines reality, and influences vision" (p. 74). For social work assessments, the authors suggested paying attention to such features as population characteristics (demographics, distribution, settlement patterns), physical characteristics of the locality (landforms, water, resources, economic and recreational potential), climate (role in community patterns, seasonal effects), meaning of place (history, myths, shared identity, sense of community, insider and outsider perceptions), and language (supports or discourages community and connectedness).

A PERSONAL EXAMPLE FROM THE
NORTHERN CANADIAN CONTEXT

Just how important is this understanding of context to the practice of social work outside of urban centres? Here is a story from my own experience in northern Canada to help answer the question. In the late 1970s, I was working with Community Corrections for the Yukon Territorial Government. My story revolves around preparations for the celebration of Canada Day on July 1 in Whitehorse, Yukon.

One particular year, a federal department in Ottawa decided to send fireworks to the Yukon's capital city to support our Canada Day celebrations. This was no minor investment, as $14,000 worth of fireworks arrived in Whitehorse (a lot of money in those days!). The Ottawa decision-makers were attempting to be inclusive. They were sending $14,000 worth of fireworks to every provincial capital city for the coming Canada Day celebrations. Why not include the territorial capitals? This was an era when southern Canadians and politicians were beginning to become aware of the territories north of 60 degrees as participant members of the country. I am sure the fireworks' gesture was a sincere attempt to include northern Canadians in the national celebration. However, the politicians were ignorant of the local context.

On July 1 in Whitehorse, Yukon, there is almost 24-hour sunlight! You can read a book outside at midnight. Summertime fireworks may go over well in Regina or Toronto or Halifax, but they are a tremendous waste in the bright evening sky over Whitehorse. I am not saying we didn't have a great party. Everyone brought lawn chairs and beer out to the airport to watch the Canada Day fireworks, which turned out to be a great joke at Ottawa's expense. Sure, we could hear the fireworks exploding overhead and we could even see the odd puff of smoke, but none of the colour, awe, and majesty of the intended display. It was simply not relevant for us in our northern context.

I use this story often in my classes on rural and remote social work. It can be received on many levels. On the surface, there is the humour of the familiar fish-out-of-water scenario. We all had a good laugh that night in Whitehorse at the bureaucrats who thought they were doing the right thing, but who did not understand the northern summer reality of the midnight sun. The humour fades, however, when we consider that it was this same arrogant southern mindset that developed and imposed programs of social welfare, health, education, and justice on northern regions. With their southern, urban-based assumptions, these programs were often inappropriate, and sometimes insulting and damaging. What might we have done in Whitehorse that year if we had been given the $14,000 to use for planning our own local celebration with events that had meaning in our context? Such a process would involve trust and a respectful appreciation of the local context. Meaningful local expression cannot easily be determined at a distance by those unfamiliar with the setting and ignorant of the possibility that the world is different outside of their urban centre.

REMOTE PRACTICE

Dissatisfied with application of the new rural social work in isolated northern regions, in the 1980s a number of Canadian authors began to discuss northern or remote practice as a distinct specialization. Collier (1984) was the first in the group to distinguish in writing between rural and remote practice in the Canadian context, with others following soon after (McKay, 1987; Nelson, McPherson, & Kelley, 1987; Wharf, 1985; Zapf, 1985). Building on the work of Canadian geographers who were defining northern regions as distinct from southern rural agricultural areas, these authors observed that social workers in isolated northern regions were reporting stresses and conflicts that were characteristic of their remote settings, but generally unrecognized in southern practice and social work education. These concerns included such things as professional isolation, role conflicts between employer and community expectations, and complex issues related to high visibility, multiple roles, and social embeddedness.

This extension from rural social work to remote practice during the 1980s was not limited to Canadian social work, although the process was not evident in the United States (outside Alaska). The European Centre for Social Welfare Training and Research proposed a new category of "remote or isolated rural areas" to recognize their difference from rural agricultural regions within the influence of urban centres (Maclouf & Lion, 1984, p. 8). Commonalities of practice issues across Scandinavia led Lindholm (1988) to call for a search for "a separate identity for Nordic social work and its training" (p. 11). Ribes (1985) also recognized northern Scandinavia as an isolated region with similarities to northern Canada. From Australia came calls for context-based models of practice to recognize local history, politics, and the environment in rural and remote outback regions (Rosenman, 1980; Cheers, 1985; Cheers & Taylor, 2005).

A note on terminology is in order at this point. International authors tend to use the term "remote practice" rather than "northern practice" because it is the extreme isolation factor and not the compass direction that constitutes the defining characteristic of the regions under study. In the Canadian context, the commonly preferred term is still "northern practice," but this does not translate well internationally as not all hinterland areas are located in the northern regions of their countries.

Since the groundwork was laid in the 1980s, the knowledge base for northern or remote social work practice has continued to develop. An entire special issue of *The Northern Review* was devoted to northern social work practice and education (Wharf, 1991). Compilations of northern practice and education experiences were assembled from northern Manitoba (Tobin & Walmsley, 1992) and northern Alberta (Feehan & Hannis, 1993). Lakehead University's Centre for Northern Studies published the Northern Social Work Series, with volumes on northern practice (Delaney & Brownlee, 1995), northern issues (Delaney, Brownlee, &

Zapf, 1996), northern strategies (Brownlee, Delaney, & Graham, 1997), northern community work (Delaney, Brownlee, & Sellick, 1999), and northern organizations (Brownlee, Sellick, & Delaney, 2001), and a final summary volume (Delaney & Brownlee, 2009).

The new literature identified and explored various features and aspects of remote social work practice. A diverse range of remote regional contexts was identified (including outport fishing villages, Inuit settlements, mining communities, logging towns, reserves, and government centres), each with its own unique features of geography, climate, history, culture, and economic development. Many fields of practice (such as family violence, substance abuse, corrections, HIV/AIDS, and child welfare) were also examined for necessary modifications in remote regions. Remote practice was presented in the light of larger social movements such as decolonization and empowerment. Articles on staffing issues considered Indigenous workers and natural helping systems in isolated areas. Discussions of appropriate remote practice models explored such approaches as generalist practice, interdisciplinary work, context-sensitive practice, community partnerships, and traditional healing practices. Throughout much of this remote/northern practice discussion, there were suggestions of a different world view or metaphor in northern and remote regions, a different relationship between people and the natural environment.

A SENSE OF PLACE

Throughout this overview of rural and remote social work, there have been emphases that were not apparent in the mainstream social work literature. Central to rural and remote practice is a notion of context, of locality, of place, and its powerful implications for human identity, activity, and problem-solving. In rural settings, a shared history and lifestyle leads to a rural identity that is rooted in a sense of belonging and a profound attachment to place (Cheers, 1998; Collier, 1993; Ginsberg, 1998; Johnson, 1997; McKenzie, 2001; Stuart, 2004; Watkins, 2004; Zapf, 2002).

How does this rural notion of "place" affect the practice of social work? I think Brian Cheers expressed this very well in his keynote address to the International Rural Human Services Conference in Halifax in 2003:

> I, *and you*, bring to the conversation my, *and your*, particular place. It is the only one there is. It is an open dynamic mosaic; a place where lives, livelihoods, environment, culture, and governance meet; a place where community, services, policy, and professional narratives intersect. Things happen in places.
>
> But I haven't come here dragging my place behind me—kicking and barking like some bewildered cattle dog on a sheep station. It lives through me. I *make* the space I live in *my place* by giving it meaning as I go about my daily living. The rural practitioner does not sit outside, mysteriously materializing in the space of the community to, just as mysteriously, disappear back to some well-ordered, comfortable, well-to-do planet of professionals when their day's work is done.
>
> If my place is unique, then so, too, is my practice. I invent it as I go along. I don't *do* my practice—I *make* it in places. (Cheers, 2004, p. 9)

A new perspective begins to arise from the centrality of this notion of "place." Social work is not something created elsewhere and then done or imposed on rural or remote areas. It is created or made in each place. Nelson and McPherson (2004) put forward a model of contextual fluidity that returns social work to "its origins of person inseparable from place" (p. 206). When Schmidt and Klein (2004) summarized the characteristics of northern social work practice, they also concluded that it is "heavily influenced by the geography of place" (p. 242). Schmidt (2005) went on to state this even more strongly with his assertion that northern social work is a form of practice that "has the environment at its core. The geography of the North greatly influences how and where people live, which in turn creates the structures and constraints of northern social work practice" (p. 18). Place is rooted in the physical environment, and it is central to rural and remote social work practice.

Collier (2006) concluded the third edition of his rural practice text with a call to transform our "rational-intentional" helping processes "into better ones which can serve humankind with equality and justice, while at the same time not damaging the foundation of resources upon which we all perch" (p. 102). Here is a balance between conventional social work processes and respect for the natural environment. There is also a suggestion here that rationality may not be sufficient for this proposed new understanding. I don't interpret Collier's suggestion to mean that social work must become irrational; instead, I see the direction as moving beyond our conventional rational and controlling approaches if we are to come to a new understanding of the relationship between people and the natural world. Some have already gone past this notion of doing no environmental damage to advocate for an active sense of responsibility for a healthy physical environment, using language that attempts to push past rationality to incorporate at least a sense of stewardship if not an outright spiritual connection. In the entry on "Remote Practice" in the new *Encyclopedia of Canadian Social Work*, I make specific reference to "an underlying worldview that emphasizes stewardship rather than exploitation" (Zapf, 2005a, p. 322). Using more inspirational language, Coates (2003) proclaimed an opportunity for social work "to establish itself as a pillar in the foundation of a new society in which humans rejoin the rest of nature in supporting the unfolding of creation" (p. 159). These notions of spiritual connections and responsibilities bring us to the next chapter.

CHAPTER FIVE

ENVIRONMENT AS SACRED

Spirituality, Deep Ecology, and Aboriginal Perspectives

The Americas are an ensouled and enchanted geography, and the relationship of Indian people to this geography embodies a "theology of place," reflecting the very essence of what may be called spiritual ecology. American Indians' traditional relationship to and participation with the landscape includes not only the land itself but the way in which they have perceived themselves and all else.

(CAJETE, 1999, P. 3)

The previous chapter concluded with indications from a rural/remote social work perspective of a spiritual connection with the natural environment. What might it mean to social work to suggest a spiritual relationship with the environment? Can social work incorporate spirituality in any meaningful way without the environment? This chapter explores a relatively recent renewal of interest in spirituality within the mainstream social work profession and literature. Related perspectives from deep ecology and Aboriginal social work are also examined for their understandings of the relationship between people and the natural world.

Spirituality and Social Work

Social work's renewed interest in spirituality has been attributed to "a longing for profound and meaningful connections to each other, to ourselves, and to something greater than ourselves" (Drouin, 2002, p. 34) that has arisen because the Western mindset of individualism and materialism has ruined the environment and destroyed community. Drouin (2002) observed a "growing spiritual longing" in social work practitioners, clients, and Western society as a whole (p. 36).

Issues of spirituality have followed a similar route to recognition within the social work profession as we saw with rural practice. The formation of a national organization leads to annual conferences and proceedings. Collections of readings and a few authored books are published, which begin to establish a theory base and consider applications to practice. Mainstream social work journals publish occasional related articles, with some journals devoting entire special issues to the subject.

In the United States, the American Society for Spirituality and Social Work (SSSW) was founded in 1990 with Dr. Ed Canda as its first director. SSSW maintains a website with links to the annual national conference, regional events, and publications. The *Journal of Religion, Spirituality, and Social Work* provides a regular forum for articles and debate. One can also find occasional articles on spirituality in mainstream social work journals with broad circulation or those with more specialized applications (e.g., the journal *Social Work and Christianity*). Since the mid 1990s, a number of American books have been published exploring spirituality and the helping professions. These include *Spirituality in Social Work Practice* (Bullis, 1996), *Spirituality in Social Work: New Directions* (Canda, 1998), *Spiritual Diversity in Social Work Practice: The Heart of Helping* (Canda & Furman, 1999), *Transpersonal Perspectives on Spirituality in Social Work* (Canda & Smith, 2001), *Explorations in Counseling and Spirituality: Philosophical, Practical, and Personal Reflections* (Faiver, Ingersoll, O'Brien, & McNally, 2001), *Integrating Religion and Spirituality into Counseling: A Comprehensive*

Approach (Wiggins Frame, 2003), and *Spiritually Oriented Social Work Practice* (Dezerotes, 2006).

Development in Canada occurred a little later with the inaugural conference of the Canadian Society for Spirituality and Social Work (CSSSW) in Toronto in 2002. Annual conferences have led to proceedings that are available through the CSSSW website, as are special issues of the e-journals *Critical Social Work* (volumes 6[2] and 7[1]) and *Currents: New Scholarship in the Human Services* (volumes 1[1], 2[2], and 3[1]). A recent book, *Canadian Social Work and Spirituality: Current Readings and Approaches* (Coates, Graham, Swartzentruber, & Ouellette, 2007), consolidates much of the Canadian material.

Rice (2002) observed that "in Australia, the issue of spirituality and social work is slow to warm up" (p. 303). While there is an active Australian Association of Spiritual Care and Pastoral Counselling, there does not appear to be a formal organization dedicated specifically to social work and spiritual issues at this time. As we saw in the previous chapter, however, spiritual issues (particularly involving connections with the land) have been central in Australian writings about rural, remote, and Aboriginal social work. *Spirited Practices: Spirituality and the Helping Professions* (Gale, Bolzan, & McRae-McMahon, 2007) brings together recent Australian readings on the subject.

Overall, there is evidence of some optimism about an apparent escalation of international connections regarding spirituality and social work, in particular "the Canadian–U.S. collaboration" (Canda, 2008, p. 33). Since 2006, the American SSSW and Canadian CSSSW have met for an annual spring North American Conference on Spirituality and Social Work.

SPIRITUALITY AND THE ENVIRONMENT: SMOKE AND MIRRORS AGAIN?

Most authors who have addressed spirituality and social work begin by making a distinction between spirituality and religion (Canda, 2008;

Zapf, 2007). The defining characteristic of religion appears to be some observable expression of a belief system through prescribed activity or ceremony. Gilbert (2000) spoke of religion as the "visible expression of beliefs" (p. 68), while Cascio (1999) regarded religion as "belief or faith ... [and] the expression of these tenets in an institutional manner" (p. 130). Spirituality, on the other hand, is presented as a much broader concept. Much of the literature makes reference to Canda's (1988) influential definition of spirituality as "the human quest for personal meaning and mutually fulfilling relationships among people, the non human environment, and, for some, God" (p. 243).

This foundational definition clearly includes relationships between people and the natural environment, yet I find in the literature that this component quickly disappears from most discussion of spirituality and social work. The broad concept of spirituality as a meaningful connection with the universe narrows significantly to become merely a quality or characteristic of the individual (Zapf, 2005c). We could be witnessing an encore of the smoke and mirrors illusion from Chapter 1, whereby the broad notion of *environment* was transformed into the narrower *social environment* with insubstantial explanation or no explanation at all. With regard to spirituality and social work, it appears to me that Canda's original broad notion of spirituality (as a meaningful connection through fulfilling relationships between people and the natural world) has been narrowed over time to simply an internal quality of the individual, again without adequate explanation. Consider the following examples from the social work literature.

Tolliver (1997) concentrated on "the spiritual dimension of the self" (p. 483), a perspective supported by Drouin's (2002) assertion that "spirituality always begins by exploring one's interior life" (p. 38). Spirituality has been described as "the divine essence of the individual" (Carroll, 1997, p. 29) and "an innate human quality" (Faiver et al., 2001, p. 2). Csiernik and Adams (2003) presented spirituality as a "foundation of client and personal wellness" (p, 65), and Sermabeikian (1994) offered a "theoretical framework for understanding spirituality within the individual" (p. 178). Cornett (1992) saw spirituality as "a legitimate clinical focus," arguing that the "psychosocial perspective," which became the "biopsychosocial perspective," must now become a "biopsychosocialspiritual model" (p. 102). Ortiz and Smith (1999) similarly advocated a "social–cultural–spiritual context" (p. 318) for social work interventions. Inclusion of spiritual problems in the fourth edition of the *Diagnostic and Statistical Manual of Mental Disorders* (American Psychiatric Association, 1994) provided an impetus for social workers to consider a client's spiritual values as a legitimate focus for clinical intervention (Foley 2001; Jacobs, 1997), yet issues of spirituality failed to become part of the training for social workers and counsellors (Lysack, 2003).

Ortiz and Smith (1999) conducted a content analysis of the term "spirituality" in the social work literature and found "the common themes of interconnectedness between self, others, and sense of ultimacy as well as the individual's need for generativity and inner meaning" (p. 309); none of these themes mentioned connection with place or the physical environment. Carroll (2001) presented two dimensions of spirituality: a vertical dimension or "relationship with the transcendent" and a horizontal dimension or "all other relationships—with self, others, and the environment" (p. 7). Later in the very same paragraph, however, she refers to this horizontal dimension as "the kind and quality of one's relationships with self and others, to well-being in relation to self and others, and to a sense of life purpose and satisfaction." What happened to relationships with the environment? They disappeared without explanation—smoke and mirrors again! The natural world was lost as spirituality focused only on the individual and social relations.

SPIRITUAL INTERVENTIONS AND THE ENVIRONMENT

What about the practice area? How has spirituality been incorporated into social work practice? Bullis (1996) offered a list of 25 spiritual interventions, ranging from the sharing of religious books to the practice of exorcism. Faiver et al. (2001) discussed spiritual problems experienced by clients and put forward their own list of 13 spiritual interventions. Wiggins Frame (2003) also presented a series of "explicit religious and spiritual strategies for counseling" (p. 183), ranging from prayer and meditation through to spiritual journalling, forgiveness, repentance, and surrender. None of these spiritual interventions involved the natural environment. Canda and Furman (1999) acknowledged that a rapport with the natural environment can be helpful for clients, but the "beauty and inspiring qualities of nature" were relegated to the margins as attending to "the aesthetics of helping" (pp. 192–193).

Hodge and Kaopua (2005) argued for the importance of spiritual assessment and briefly reviewed six qualitative instruments for conducting such assessments. Again, these instruments appeared to focus on the individual client's spiritual beliefs and practices. I found no mention of the physical environment or non-human world, although the "Spiritual Eco-Map" did include "that portion of clients' spiritual stories that exists in present space" to "emphasize clients' existing relationships to spiritual assets in their here-and-now environments" (p. 4). Elements of the physical environment might be included here, but only as background for the client's spiritual stories and issues.

Others have attempted to categorize spirituality as a variable that is subject to social work processes of manipulation and evaluation. Russel (1998) called for research on the "efficacy of spiritually derived interventions" (p. 27), while Hodge (2001) found evidence in case studies documenting "spirituality's salience in a wide variety of areas" and concluded that spirituality can be a "significant variable in recovery" (p. 203). With regard to social work education, Cascio (1998) argued that the addition of spiritual content "need not force significant restructuring of existing classes ... the spiritual dimension can thus be presented as an additional aspect to be considered" (p. 530). Sadly, it appears to me that spirituality in these accounts has been diminished to a mere option for clinical practice, another intervention technique added to the worker's repertoire, another variable to be considered for outcome research. Perhaps the most dramatic instance of this perspective is Kasiram's (1998) reference to "the use of God" as a resource in therapy (p. 172)! Here we have the ultimate variable, the trumping strategy. (I would not want to be the worker who concludes in the case file that the use of God was not helpful in this instance, or that the use of God was less effective than medication.)

A recent mainstream generalist practice text by Zastrow (2007) included a chapter on "Spirituality and Religion in Social Work Practice." This recognition of spirituality as central to our training and practice held initial promise, but I was disappointed to find that the narrow, person-centred definition of spirituality appeared to be assumed. The whole discussion was presented under the umbrella of "culturally sensitive practice" (p. 365), with an emphasis on respect and appreciation for beliefs that are different from one's own. Key components of spirituality were identified to include "the personal search for meaning in life, having a sense of identity, and having a value system" (p. 366). Questions were put forward for conducting assessments that included "spiritual aspects of clients" (p. 370), as well as guidelines for choosing appropriate spiritual interventions with individual clients. Once again, spirituality was conceptualized as a characteristic of the individual with no mention of connections with the physical environment or non-human world.

Not everyone is ignoring connections with the natural world in their explorations of spirituality and social work. A decade after contributing the foundation definition, Canda (1998) called for social work to revisit the person-in-environment concept "in a dramatic way" because the person is "not separable" (p. 103) from the natural environment. The next year,

Canda and Furman (1999) further challenged the profession to reconsider "what is the whole person and what is the whole environment?" (p. 194). Foley (2001) presented the new spirituality as holistic, a "new awareness of the human within the intrinsic dynamics of the earth itself" (p. 356). Ingersoll (2001) presented an assessment tool called the Spiritual Wellness Inventory that included a component of connections with the natural environment. Another mainstream text, *Human Behavior and the Social Environment: Macro Level* (van Wormer et al., 2007), discussed in Chapter 3 for its material on the physical environment, concluded with a chapter on the religious/spiritual environment for human behaviour. Featured there is a reading on "Earth as a Source of Spirit" (Sheridan, 2007), which argues for connection with the natural world as a place of nurturance, healing, and renewal. Social workers are encouraged to consider employing nature as "a healing modality" (p. 270) because "Earth is truly a source of spirit in all its manifestations" (p. 272).

In spite of these few encouraging accounts, however, I would argue that the call to revisit person-in-environment has mostly been ignored as the profession appears content for now with its limited understanding of spirituality as a human quality or social characteristic, excluding any relationship with the non-human world. Most pleas to reconsider person-in-environment have only resulted in directing more attention to either the person or the environment, while continuing to ignore the relationship. We are seduced by the nouns, a situation that will be explored further in Chapter 8.

Deep Ecology

The developing approach of deep ecology has been described as a "small but growing movement within social work" (van Wormer et al., 2007, p. 23). Consistent with Canda's (1988) broad definition of spirituality, deep ecology clearly rejects divisions between the human and non-human worlds, and suggests instead that human identity derives from an ecological consciousness:

> [A] moving away from a view of person-in-environment to one of self as part of a "relational total-field" ... rather than experiencing ourselves as separate from our environment and existing *in* it, we begin to cultivate the insight that we are *with* our environment. (Besthorn, 2001, p. 31)

According to Devall (1980), "deep ecology is not an attempt to add one more ideology in the crowded field of modern ideologies. Deep ecologists are questioning for ways to liberate and cultivate the ecological consciousness" (p. 317). Rosenhek (2006) put it simply: "In a nutshell, the deep ecology movement reminds us that we are from the Earth, of the Earth and not separate from it" (p. 91).

The label "deep ecology" can be confusing when first encountered. Depth of what? Naess (1973) first wrote of a deep and long-range ecological movement, which he distinguished from superficial or shallow ecology that focused on more immediate problems (often economic) caused by temporary and local environmental issues. Deep ecology was seen to be more concerned with the philosophy of the profound relationship between humans and nature, a philosophy Naess (1973) called "ecosophy." As explained by van Wormer et al. (2007):

> In deep ecology, the concept of environment includes all human and nonhuman beings, processes, things, and systems in the total planetary ecology and, indeed, the larger cosmic ecology.... Deep ecology is called "deep" because it helps people to understand connections with the world in profound and self-transforming ways. Deep ecology leads to a re-envisioning of both person (as total person) and environment (as total ecology). In this sense, person and environment are

codetermining concepts, because person cannot exist without environment. (p. 243)

Deep ecology promotes harmony and connection among all forms of being, a mutual dependence rather than human domination of the natural world for economic gain. Diverse ecosystems have intrinsic value beyond their economic utility for extractable resources. Because we are aware as humans of our considerable impact on the natural environment, we have a special responsibility to care for the Earth (LaChappelle, 1988).

Deep ecologists advocate a broader comprehensive notion of self as the "ecological self" (Besthorn, 2002; Devall, 1995; Lysack, 2007; Mathews, 1991; Seed, Macy, Fleming, & Naess, 1988) or simply as "self" (Drengson & Inoue, 1995). This ecological self moves beyond individual ego to consider the larger ecosystem as "self," a perspective of "eco-consciousness" (Dezerotes, 2006, p. 34). Every component, human and non-human, has intrinsic value and is related to all other components. Morito (2002) explained:

> As we see ourselves as parts or aspects of a larger Self, change in the egoistic conception of self begins to take place. We then become sensitized to the needs of other creatures, who form the basis for the extended Self. Recognizing other ecosystem inhabitants as part of the extended person is a way of seeing the ecosystem as a person to whom respect is owing. (p. 184)

Organizational structures related to deep ecology and social work are still relatively new. *The Trumpeter: Journal of Ecosophy* has been in publication since 1983, serving as a forum for discussion and development of the movement. Founded in 2000, the Global Alliance for a Deep Ecological Social Work (GADESW) held a 2001 symposium in Washington, DC, entitled "Deepening Earth Consciousness in Social Work." There are plans for future international symposia and a journal. At this time, GADESW's website serves as a forum for papers, discussion, links, and announcements related to deep ecology and social work.

What does this deep ecological perspective mean for social work? From the perspective of deep ecology, "social work practice needs to address the problems that arise from excessive and destructive human interference with nature" (van Wormer et al., 2007, p. 249). Individuals, families, and communities of people may still be the focus of many of our interventions, but we cannot exclude all consideration of the other life forms and the natural world around us. That would be "shallow ecological activism [which] leads to the absurd result that we inadvertently destroy the sustaining natural environment while trying to help people live better" (p. 248). As explained by Ungar (2002), "diversity, complexity, and symbiosis are in our own best interest" (p. 486).

Ungar (2003) went on to envision an ecological society based on the natural order. Promotion of mutual respect and holism in such a society would discourage forms of oppression and exploitation that are common in our present society. Others have also made this connection between oppression of people and oppression of nature, between social injustice and environmental injustice. Besthorn and McMillen (2002), for example, integrated principles of eco-feminism with deep ecology. They called for social work as a profession to broaden its understanding of our interconnections with the natural world, and to develop an awareness of new professional roles we may be called upon to play.

A new name for the profession as we begin to assume these new environmental roles was proposed by Ungar (2003). He argued that "those who perform 'work' upon others" (p. 20) may not be well suited to practice from ecological principles. He suggested we become "professional social ecologists" rather than social workers because "a change of name to the profession of social ecology would reposition social work practitioners on the leading edge of progressive ecological thought" (p. 5). In his guide to postmodern social work practice, Ungar (2004) began with "positioning" and the social worker's role as an "inside outsider" (p. 490). He directed attention to consideration of geographic characteristics shared by the social worker and those being helped, with a special focus on the physical location where the helping work takes place.

ABORIGINAL SOCIAL WORK: TRADITIONAL KNOWLEDGE, THE LAND, AND IDENTITY

There is evidence of a new literature beginning to develop around the theme of Aboriginal social work. Contributions include books such as *No Place for Violence: Canadian Aboriginal Alternatives* (Proulx & Perrault, 2000), *Seeking Mino-Pimatisiwin: An Aboriginal Approach to Helping* (Hart, 2002), and *Protecting Aboriginal Children* (Walmsley, 2005). Many chapters in social work collections have also addressed Aboriginal practice; for example, Meawasige's (1995) chapter on the healing circle, Bruyere's (1999) chapter on the decolonization wheel, Baskin's (2005) chapter on healing through building healthy relationships, and Longclaws' (2005) chapter comparing social work and a medicine-wheel framework for healing. Since it began publication in 1997, the *Native Social Work Journal* has served as a forum for articles on Aboriginal social work practice, research, values, and programs.

Note here that when I refer to a "new" or "developing" literature on Aboriginal social work, it is only the written literature that is relatively new. Aboriginal healing approaches have a rich and long oral tradition that predates Western social work by thousands of years. The new development is the written contribution of some of these teachings for the benefit of those of us who may learn primarily through the written word. Some articles in this new literature use terms such as "Native social work," "Indian social work," or "Indigenous social work." Baskin (2005) deliberately chose to use "Aboriginal" in her writing because, for her, "this term includes all of us—status, non-status, Metis, those of mixed blood, and Inuit. It is important to me that I be inclusive" (p. 171). Persuaded by this sense of inclusiveness, I also use the term "Aboriginal" in this discussion.

Another point of clarification is necessary before proceeding. I briefly positioned myself in relation to this material by introducing my voice at the beginning of this book. Although I have spent many years working with and learning from Aboriginal colleagues, I am not Aboriginal. I lack the experience or the authority to attempt to explain the Aboriginal world view and healing practices. It would be disrespectful and wrong for me to try to do this. Fortunately, some Aboriginal authors have risked sharing concepts and values from their healing traditions so that the rest of us can begin the process of coming to understand. In his comparison of ecological social work practice with Anishnaabe healing principles, Longclaws (2005) concluded that "the Anishnaabe medicine wheel is not a model of social work, but the teachings can provide social workers with useful tools" (p. 367). It is in this spirit that I present the following discussion, relying heavily on the original words of Aboriginal writers. The developing literature on Aboriginal social work offers many opportunities to the mainstream profession, particularly when it comes to understanding the natural environment.

What is Aboriginal social work? The term itself appears to have a range of applications. Some have written about the delivery of mainstream programs in Aboriginal communities; some have written about Aboriginal social workers practicing in Aboriginal communities; some have written about Aboriginal social workers practicing in any setting; and some have written about traditional healers. Rather than defining Aboriginal social work in terms of who is practicing with whom, a more productive approach might be to look at the world view that underlies effective social work and healing approaches in Aboriginal communities. Some authors have attempted to include traditional knowledge or Aboriginal theory as part of the knowledge base for mainstream social work practice, but any assumption of traditional knowledge as just another content area disguises a fundamental difference in world view (Zapf, 1999). How does this world view differ from the person-in-environment perspective of Western social work? The essence of the difference was expressed effectively by Morrissette, McKenzie, & Morrissette (1993, p. 93): "While Aboriginal people do not embrace a single philosophy, there are fundamental differences between the dominant Euro-Canadian and traditional Aboriginal societies, and these have their roots in differing perceptions of one's relationship with the universe and the Creator."

Aboriginal perspectives on the relationship between people and the universe have been explored and expressed through science/knowledge systems with English-langauge labels such as "Indigenous science" (Cajete, 2000; Colorado, 1991), "traditional knowledge" (Wenzel, 1999), and "Native American science" (Little Bear, 2000). Such Indigenous science is concerned with "contextual and relational knowledge" (Cajete, 2000, p. 98), described by Wenzel (1999) as "knowledge and values which have been acquired through experience, observation, and from the land or from spiritual teachings, and handed down from one generation to another" (p. 113). Cajete (2000) offered a working definition of Indigenous science as "that body of knowledge that is unique to a group of people and that has served to sustain those people through generations of living within a distinct bio-region" (p. 268). Prominent in these definitions is an element of the land, of the natural environment.

Sahtouris (1992) elaborated how the traditional way of knowing "is not a science that stands apart from nature to look at it objectively; it does not eliminate the sacred, but integrates it. It fosters dialogue between humans and the rest of nature" (p. 4). A strong spiritual foundation is the basis of Indigenous science, with a keen emphasis on the land. Connection to the land "suffuses the tribal world" (Coates, 2004, p. 47).

Indigenous societies have traditionally been built around a symbiotic relationship with their homelands.... The indigenous societies identified closely with their specific setting and developed cultural forms, habits, movements, and harvesting activities which permitted them to sustain life in a particular ecological niche. (Coates, 2004, pp. 48–49)

The entire natural order is reflected in each geographic place or region; by seeking and coming to understand their place in a given local system, people may also find their place in the cosmos. As Little Bear (2000) explained:

The land is a very important referent in the Native American mind. Events, patterns, cycles, and happenings occur at certain places. From a human point of view, patterns, cycles, and happenings are readily observed on and from the land. Animal migrations, cycles of plant life, seasons, and cosmic movements are detected from particular spatial locations; hence, medicine wheels and other sacred observatory sites. Each tribal territory has its sacred sites, and its particular environmental and ecological combinations resulting in particular relational networks. All of this happens on the Earth; hence, the sacredness of the Earth in the Native American mind. The Earth is so sacred that it is referred to as "Mother," the source of life. (p. xi)

When the inhabitants of a region have been there for many generations, their identity incorporates that place and their relationship to it. Through this process, Aboriginal cultural identities become "tied so directly to the land and concepts of place" (McCormack, 1998, p. 28). Graveline (1998) talked about a direct link between "geographical space and worldview" (p. 19). Cajete (2000) called this link "geopsyche" (p. 187), whereby people assume traits of a particular place they have occupied for a long time and the place assumes human traits, in a continual process of co-creation. As expressed by Spretnak (1991, p. 91), "A people rooted in the land over time have exchanged their tears, their breath, their bones, all of their elements—oxygen, carbon, nitrogen, hydrogen, phosphorus, sulfur, all the rest—with their habitat many times over. *Here nature knows us.*"

With specific reference to the Western Apache, Basso (1996) elaborated on this same interconnectedness of place, spirit, and self:

> As Apache men and women set about drinking from places—as they acquire knowledge of their natural surroundings, commit it to permanent memory, and apply it productively to the workings of their minds—they show by their actions that their surroundings live in them. Like their ancestors before them, they display by word and deed that beyond the visible reality of place lies a moral reality which they themselves have come to embody. And whether or not they finally succeed in becoming fully wise, it is this interior landscape—this landscape of the moral imagination—that most deeply influences their vital sense of place, and also, I believe, their unshakable sense of self ... selfhood and placehood are completely intertwined. (p. 146)

Describing the northern bush environment where he was born and raised, Harper (2004) told of being surrounded by lakes, rivers, trees, and wild animals. As a child, he had no concept of being Canadian and had never heard of the country. "All I knew was that I belonged to the land in an integral and eternal way" (p. 163).

I am reminded here of a story that I heard from a woman at a recent workshop. From the audience, she explained that she was Dene from a region of the Northwest Territories (unfortunately, I do not know her name). She told us that her grandmother had taught her about the land. She had asked her grandmother what would happen if all the Dene disappeared from the land. The grandmother said that it would not matter if the people who came next were red, white, yellow, black, or purple. The land would teach them to be Dene.

The spirituality of traditional knowledge appears profoundly connected with land and place. Holst (1997) observed that "a spiritual landscape exists within the physical landscape" (p. 150). According to Cajete (2000):

The life of the Indigenous community is interdependent with the living
communities in the surrounding natural environment.... Indigenous
cultures are really extensions of the story of the natural community of
a place.... The geographical and structural orientations of Indigenous
communities to their natural place and the cosmos reflected a communal consciousness that extended to and included the natural world in
an intimate and mutually reciprocal relationship. (pp. 94–95)

It seems that some places within this living landscape can hold more
power or spiritual energy than others, and these are often called sacred
sites. In Walker's (1998) terms, sacred sites are "portals to the sacred" or
"geographic points that permit direct access to the embedded sacredness
in nature" (p. 6). Peat (1994) also explored this notion of natural energy
connected closely with a geographic sense of place:

Unlike Western science, the importance of the landscape, and specific places in it, is a characteristic of all Indigenous science.... Within
Indigenous science there is an association of spirit or energy with particular places, and it is important to visit these places and carry out cer-
emonies there.... This idea of the significance of place and the energies associated with it is common to Indigenous sciences all over the
world.... Western science does not appear to have a corresponding
concept. (pp. 265–267)

Aboriginal healing practices derive from this profound connection
with the land, a living physical environment that is a source of energy
and knowledge (Colorado, 1991; Ellerby, 2000; Hart, 2002, 2006; Neil
& Smith, 1998; Proulx & Perrault, 2000; Rajotte, 1998). Rather than a
passive background, the environment is a "sensate conscious entity suf-
fused with spiritual powers" (Spretnak, 1991, p. 90). With specific reference to social work, Hart (1996) explained the Western and Aboriginal
approaches in this way:

Western models of healing separate and detach individuals from their social, physical and spiritual environments, isolating "patients" for treatment purposes and then re-introducing them into the world. Traditional healers are concerned with balancing emotional, physical, mental, spiritual aspects of people, the environment, and the spirit world. (p. 63)

Why do Western society and the social work profession have such difficulty understanding the Aboriginal perspective of spiritual connection with the land? Deloria (1999) suggested that Western society could "attribute to the landscape only the aesthetic and not the sacred perspective" (p. 257) because we tend to relate to the physical environment through technology. Cajete (1994) similarly lamented our "cosmological disconnection from the natural world" (p. 25). Cummins and Whiteduck (1998) postulated a number of cultural traits that inhibit Western society's ability to recognize sacred sites or appreciate their power, traits including "an emphasis upon the rights of the individual to pursue economic gain as opposed to the communal good, and a belief that humankind is separate from, as opposed to a part of, the natural world" (p. 12).

As we saw in Chapter 3, such separation of humans from the natural world has resulted in a Western concept of land as defined space that is subject to individual or collective ownership. In contrast, "tribal land use was typically understood in terms of stewardship and responsibility, rather than ownership" (Coates, 2004, p. 50). According to Chatwin (1987), Aboriginal Australians "could not imagine territory as a block of land hemmed in by frontiers"; instead, they understood territory as a network of interlocking "lines" or "ways through" (p. 56) the land. Survival involved moving through the landscape, not as aimless wanderers, but with intention according to the patterns of exchange with other people and nature. The Australian land was mapped according to "songlines," stretches of land defined by songs and verses passed regionally from generation to generation.

In Western urban society, we tend to view the physical environment as separate from ourselves, as an objective thing, as a commodity to be developed or traded or wasted or exploited, as an economic unit, as property. The dominant Western urban culture has been described as "hostile to nature" (Spretnak, 1991, p. 102) and antagonistic to any concept of personhood beyond individualism. Writings in eco-feminism have explored connections between the oppression of women and oppression of the environment (Booth, 1997; Warren, 1998). Parallels go beyond simple neglect or lack of respect; women and the natural environment have both been victimized by domination, aggression, overt oppression, and abuse from patriarchal groups and societies. Aboriginal belief systems have been acknowledged by eco-feminists as a promising base for environmental awareness in the larger society (Booth & Jacobs, 1993).

The foundation metaphor of Aboriginal traditional knowledge has been characterized in the literature as a perspective of "I am I and the Environment," (Ortega y Gasset, 1985). Suopajarvi (1998) explained it this way: "I'm not in the place but the place is in me" (p. 3), similar to Cajete's (2000) observation that "we are the universe and the universe is us" (p. 60). Spretnak (1991) asserted that "here nature knows us" (p. 91). Chatwin (1987) related a conversation with an Aboriginal Australian who explained how the definition of a person's "own country" was "the place in which I do not have to ask" (p. 56). Such Aboriginal expressions of "world-image identity" contrast sharply with Western culture's "self-image concepts" (Stairs & Wenzel, 1992).

In the introduction to *The Sacred Balance: Rediscovering Our Place in Nature*, Suzuki (2002) described a moment when he realized the power of language in presenting environmental issues. He recalled developing a document to present at the 1992 Earth Summit in Rio de Janeiro:

> When I was working on the first draft, I tried writing "We are made up of molecules from the air, water and soil," but this sounded like a

scientific treatise and didn't convey the simple truth of our relationship with Earth in an emotional way. After spending several days pondering the lines, I suddenly thought, "We *are* the air, we *are* the water, we *are* the earth, we *are* the Sun."

With this realization, I also saw that environmentalists like me had been framing the issue improperly. There is no environment "out there" that is separate from us. We can't manage our impact on the environment if we *are* our surroundings. Indigenous people are absolutely correct; we are born of the earth and constructed from the four sacred elements of earth, air, fire and water. (Hindus list these four and add a fifth element, space.)

Once I had finally understood the truth of these ancient wisdoms, I also realized that we are intimately fused to our surroundings and the notion of separateness or isolation is an illusion. (pp. 7–8)

Harper (2004) explained the Aboriginal relationship with the land in this way:

Land is very sacred to us. It is essential to our existence, our philosophy, our way of life. We live on the land, we belong to it, and we return to it when we die.... Our relationship with this land is a responsibility that is not in our power to extinguish. (p. 166)

Such spiritual interconnection and mutual responsibility between self and nature moves us far beyond social work's current notion of person-in-environment. Deep ecology brought us to a perspective that we are *with* our environment rather than *in* it. Aboriginal world views offer an understanding of person *as* environment, humans and the natural world understood as two inseparable expressions of the same Creation. The implications of these prepositions and relationships are explored further in Chapter 8.

CHAPTER SIX

INTERNATIONAL SOCIAL WORK

Environment, Development, and Sustainability

Certain social issues transcend nation-states,
including environmental changes such as destruction of
the ozone layer, global warming, and deforestation.
(SOWERS & ROWE, 2007, P. 12)

Acknowledging that the voices of international social work may be closer to the core of the profession than the voices of rural/remote practice or spirituality, I still consider them to be coming from the margins of the mainstream profession in North America. Turner (2005d) argued that international social work, owing to "its very rapid development in recent years" (p. 198), has attained the status of a social work method or approach, but few others have accorded it this central position. International social work content in the curricula of schools of social work is typically located in option/elective courses or in modules as supplements or add-ons to core courses. Introductory social work textbooks may include a discussion or chapter on international social work somewhere near the end. Yet international social work does have its own long-standing organizations and literature. Just what is international social work and what does it have to say about the natural environment?

International Social Work and Global Environmental Citizenship

Not surprisingly, the literature offers a range of definitions of international social work with only broad areas of consensus. The approach is often labelled with various combinations and permutations of "global," "international," "comparative," "social work," "social welfare," and "social development." Thus, one author may refer to "international social welfare" (James, 2005) while others explore "international social work" (Healy, 2001; Hokenstad, Khinduka, & Midgley, 1992), "international development" (CIDA, 1987), "globalization and social work" (Drover & MacDougall, 2002), "globalization and social welfare" (Drover, 2005), or "global social work" (Sowers & Rowe, 2007). An overview of the definitions offered in the literature suggests to me several broad areas of focus for international social work: comparisons of social work approaches and programs in different countries or regions of the world; transfer of social work knowledge, models, and personnel between countries and regions; international agencies, organizations, and services that use social work approaches; and development of international social policy.

Ife (2000) challenged the common social work slogan to "think globally and act locally" by suggesting it is no longer helpful for addressing today's international issues:

> It has become necessary to think and act at both local and global levels, and to link the two. The problems of our clients are caused in part by global forces and if we are to seek adequate solutions, it is necessary to engage in global action. (p. 62)

In response to new realities of global economics and governance, new models of citizenship may be necessary to express the active roles of participants in the globalized world. Falk (1994) suggested five emerging forms of global citizenship: aspirational citizenship; citizenship of transnational affairs; citizenship of environmentalism; regional citizenship; and citizenship of activism. Using this framework, Drover (2000) explained citizenship of environmentalism as

> based on the necessity to address issues like deforestation, fossil fuel consumption, global warming, and the depletion of marine life. It is motivated by principles of sustainability and sensitivity to the natural order. It presupposes a conception of rights that transcends current generations. It attempts to replace a geopolitical agenda by an ecological agenda. (p. 33)

Latta (2007) explained environmental citizenship as a link between social justice and ecology, "a way to encapsulate and promote this powerful connection between environmental values and the formal relationships of political community" (p. 18). From this perspective, environmental citizenship can be examined for instances of oppression, marginalization, and exclusion in much the same way as we apply these concepts to social/ political citizenship. Who is excluded from decision-making regarding the environment? Do citizens of this planet have a fundamental right to a healthy and supportive physical environment? Has the natural environment itself been oppressed by human activity, prevented from full participation and expression of its life forces?

Because protection of the environment requires collective action at the global level, the notion of global environmental citizenship pushes beyond individualism and nationalism. Issues of global warming, air pollution and water supply do not fit neatly within national jurisdictions or the private property rights of individuals. Meaningful change must transcend local and national efforts. The vision of Chan and Ng (2004) for social work in the 21st century involved cultivating a sense of global citizenship that included "establishing a collective ownership for public goods such as the global environment" (p. 314). South African social activist and CEO of the World Alliance for Citizen Participation Kumi Naidoo envisioned a "global civil society" (Naidoo, 2004, p. 12) taking on the challenges of environmental destruction in an interdependent world.

Closely associated with the concept of environmental citizenship is the notion of environmental rights. While the right to a healthy environment may be assumed in many human-rights documents, Kolari (2004) pointed out that these rights are not clearly articulated and actually provide little real protection for the earth's natural environment. Rather than trying to develop some new environmental-rights instrument, however, Kolari (2004) advocates for a more effective approach involving "the development of all human rights in a manner that demonstrates that humanity is an integral part of the biosphere, that nature has an intrinsic value and that humanity has obligations towards nature" (pp. 141–142). This approach parallels Morito's (2002) distinction between thinking about ecology and thinking ecologically. The interrelationship between humans and nature should be a foundation of all human rights, and not relegated to some new and separate set of environmental rights.

DEVELOPMENT AND THE ENVIRONMENT

Lightman (2003) looked at globalization and critiqued as meaningless any attempt to separate social programs from economic activities in a treaty such as the North American Free Trade Agreement. Exploring the links between social and economic development in the international context, some have identified environmental damage as a real cost associated with such economic growth, and environmental disasters as the physical manifestation of the limits to social and economic development (James, 2005; Midgley, 1997). Modern society's emphasis on economic development has resulted in a degraded and polluted physical environment that can bite back at us.

Environmental disaster was explored by Rowe (2005), who argued that social workers must be prepared for "eco-disaster" and be ready to help clients cope with environmental calamities that destroy hope and pose existential questions. When faced with such eco-disasters, many people "simply leave a depressed area in hopes of improving their economic opportunities elsewhere" (p. 206). Is this still possible? Is there always somewhere else to go where things are better? Or could the planet itself be facing ecological disaster, leaving us with no place to go?

In a thoughtful discussion of vulnerability analysis as an aspect of disaster research, Varley (1994) presented an understanding of natural disasters as the product of interaction of social processes with the physical environment. "Although floods or earthquakes are natural processes, the disasters associated with them are not" (p. 1). Assessing human vulnerability and developing strategies to mitigate the effects of natural disasters must involve a focus on social processes and structures, in addition to an understanding of the natural forces at work (Cannon, 1994). Yet the dominant frameworks in disaster management have focused narrowly on extreme natural hazards as the problem and on technology as the solution. As expressed by Mitchell (1990), the idea of "interaction among physical risks and human responses is largely bypassed in favor of a focus solely on physical risks ... [accompanied by] expansive and optimistic assumptions about the role of natural science and engineering knowledge in the hazards policy arena" (p. 147). With regard to natural disasters, it seems we continue to think we are in control when it comes to solutions while denying our responsibility for the underlying societal and natural conditions.

An interesting notion of "environmental racism" has been introduced (Bullard, 1993, 1994; Sowers & Rowe, 2007; van Wormer, 1997; van Wormer et al., 2007) to describe how the toxic effects of environmental pollution are not experienced equally. In North America, toxic dumps, chemical plants, and incinerators are distributed disproportionately among impoverished and non-white locals; waste management and environmental laws may be selectively enforced; children from poor families are more likely to have higher exposure to environmental contaminants and subsequent health problems. There is also evidence of environmental racism on a global scale. Consider the French attempt to scrap a carrier loaded with toxic waste at a shipyard in India (Ahmed, 2006) or the use of rural China as a disposal site for American e-waste (Garber, 2007), resulting in high levels of lead poisoning in local children.

A common theme in the international social work literature is concern about the international implications of a continuing focus by North American social workers on the individual in spite of the profession's declaration of a person-in-environment perspective (Drucker, 2003). Midgley (1981) used the term "professional imperialism" to refer to this widespread influence of Western social work theory and practice around the globe. Caragata and Sanchez (2002) warned of the danger that international social work education and practice might become too grounded in Western expertise with individual therapy and clinical practice. In the developing world, they argued, human problems must be seen as more closely connected to the environment and its resources. It would appear that, from an international perspective, welfare of the environment is a major concern alongside welfare of people.

Reporting on the incorporation of environmental issues into social work practice in the United States and South Africa, Marlow and Van Rooyen (2001) found that American social workers tended to focus on personal efforts (e.g., individual conservation and recycling activities) whereas South African social workers tended to be more involved with community-based efforts (e.g., co-operative clean up of the immediate environment of a village). The authors expressed concern that Western social work has focused only on the social aspects of the environment and distanced itself from any consideration of the physical environment. They concluded that, for international social work, "the theories guiding practice need to be re-visited so that they are inclusive of the environment at the broadest level" (p. 252).

Social work authors from many parts of the world have called for the profession to push beyond the conventional narrow perspective on person-in-environment (dominant in the North American literature) and move towards a focus on community development that incorporates the natural environment. A selection of these accounts will illustrate the point.

INTERNATIONAL VOICES FOR SUSTAINABLE DEVELOPMENT

McKinnon (2005) observed that "in developing countries such as India, Indigenous social work has taken a broader approach to social and welfare considerations—the everyday physical world of individuals and the community is considered important" (p. 226). From South Africa came a similar plea for the relationship between people and the physical environment to become central to social work (Cock, 1991). Environmental concerns can no longer be left as intellectual issues for debate among the upper classes. The relationship between people and the environment affects everyone, everywhere. Speaking from Brazil, both Rodwell (1995) and Cornely and Bruno (1997) called for an expansion of the profession's notions of research and macro practice to build interventions from the local community and resources. Social action related to land development was fundamental to their vision of social work. Wint (2000) advocated for sustainable communities as the goal in Jamaica, as did Schobert and Barron (2004) in rural Haiti. From Nepal, Pandey (1998) argued for social work to tackle the issues of degradation of natural resources, especially forests, which were threatening the survival of rural populations. "Social workers are best suited and have real opportunity to create a favorable policy and program environment" (p. 352). A similar argument was offered for the Amazon region, where social work needs to move beyond remedial work following ecological disasters to a more progressive environmental-development planning approach (A. Hall, 1996).

In northern Ethiopia, Fitzgerald (1994) observed that community-based environmental programs designed "to conserve soils and halt environmental degradation" (p. 125) were not sufficient to overcome famine when they were divorced from relevant social, economic, and political concerns. From the Philippines, Quieta (2003) similarly argued that a balance with the natural environment must be central to any community-development strategies:

> The emphasis on the environment and ensuring that it is protected and sustained brings all parts of the country's community development system together.... The Philippines has joined many other countries in seeking to protect the environment and to promote sustainable economic development policies.... Social workers have played an increasingly important role by using their skills in organizing and building strong relationships to promote environmental ideals among local people. (pp. 71–72)

Most of these international accounts make reference to a vision of sustainable development. A report from the United Nations World Commission on Environment and Development (1987), often referred to as the Brundtland Report, gave meaning to this expression with its simple yet powerful definition of sustainable development as "development which meets the needs of the present without compromising the ability of future generations to meet their own needs" (p. 43). According to van Wormer (1997), social workers around the world should be "guided by sustainable social development concepts to contribute to policy decisions needed in this time of unprecedented global challenge" (p. 30). The final chapter of her book on international social work is devoted to the physical environment, more specifically "the convergence of social and environmental problems so often neglected in the social work literature"

(p. 643). Issues of environmental degradation, war, and overpopulation are examined, with the conclusion that sustainable development is a necessary paradigm to rethink and reverse the damage done by unchecked economic development. Tester (1997) similarly presented a new paradigm of community development as an environmental movement.

Applying United Nations measures of human development and ecological footprint, Guevara-Stone (2008) identified Cuba as the most sustainable nation in the world. Contributing to Cuba's remarkable energy revolution is a small army of *trabjadores sociales*, a force of "13,000 social workers [who] have visited homes, businesses, and factories around the island, replacing light bulbs, teaching people how to use their new electric appliances and spreading information on saving energy" (p. 24). These social workers are also involved with macro level conservation initiatives for the national bus system and the annual sugar cane harvest.

There is a caution here. With a postmodern tendency to collect and learn from local voices in various settings around the globe, we might easily miss the big picture, the grand narrative, the necessary international movement towards environmental sustainability. Noble (2004) reminded us that "grand narratives are still having currency in challenging the effects of neo-conservative politics. Rather than call for the negation of grand theories, social work needs to realign itself with a more, rather than less, national and global focus" (p. 302).

INTERNATIONAL SOCIAL WORK ORGANIZATIONS: GLIMPSES OF THE ENVIRONMENT

Three major organizations are generally considered to represent the interests of social work at the international level (Kimberley & Osmond, 2005; Sowers & Rowe, 2007): the International Federation of Social Workers (IFSW), the International Association of Schools of Social Work (IASSW), and the International Council on Social Welfare (ICSW). As James (2005) explained, "simply stated, the IFSW serves the social work profession, the

IASSW is the voice of social work education, and the ICSW is concerned with all aspects of social welfare" (p. 497). Founded in 1956, the IFSW currently has more than 745,000 members through national associations in 90 countries (IFSW, 2008), and produces the journal *International Social Work*. IASSW membership is comprised of some 517 faculties and schools of social work from 72 countries (IASSW, 2008). The ICSW is a federation of national committees and local organizations from more than 70 countries (ICSW, 2008).

Of particular interest to this discussion is the "international definition of social work" that was adopted by the IFSW General Meeting in Montreal in 2000 and jointly agreed by the IFSW and IASSW in Copenhagen the following year (IASSW, 2001; IFSW, 2000). Here is that international definition of social work:

> The social work profession promotes social change, problem solving in human relationships and the empowerment and liberation of people to enhance well-being. Utilising theories of human behaviour and social systems, social work intervenes at the points where people interact with their environments. Principles of human rights and social justice are fundamental to social work. (IFSW, 2000)

Given the evidence of concern from around the world for social work to consider sustainable development and balance with the natural world, I would have expected a central place for the physical environment in any international definition of social work. Can the physical environment even be found in this definition adopted by IFSW and IASSW? The definition itself consists of only three sentences. The first sentence includes reference to "social change," "human relationships," "empowerment and liberation of people," and "well-being." These are all social terms. The second sentence refers to our knowledge base of "theories of human behaviour and social systems," with no mention of any framework for understanding our relationship with the natural world that supports us. The final sentence asserts our profession's fundamental commitment to "principles of human rights and social justice"; absent is any sense of environmental justice or responsibilities.

The only place where the environment even appears in the definition is the assertion at the end of the second sentence that "social work intervenes at the points where people interact with their environments." Given that the declared purpose, knowledge base, and values are all expressed in social terms, I am initially inclined to assume that the declared focus of intervention must be the interaction between people and their social environment. Yet, there is that "s"! The definition refers to "environments" not "environment." Reference to "environments" in the plural suggests the real possibility of multiple environments, at least more than the social environment alone.

Describing the six years of work that went into formulating this definition, Hare (2004) offered this explanation for the wording regarding environments:

> The task force and the member associations ultimately agreed that the central organizing and unifying concept of social work universally was intervention at the interface of human beings and their environments, both physical and social, thereby reaffirming the thinking of previous social work theorists. (pp. 409–410)

While the physical environment may have been an important component for the IFSW task force charged with developing the definition, I am disappointed that this same physical environment was not made explicit in the final definition. I wish the IFSW/IASSW definition of social work had explicitly included the physical environment, but at least the door was opened to the possibility with the pluralization of "environment." This tiny opening, this "s," this single letter among the 318 letters in the definition (yes, I counted), may be the means for eventual expansion of the definition. Indeed, on the IFSW website, the international definition of social work is accompanied by a proviso that "social work in the 21st century is dynamic and therefore no definition should be regarded as exhaustive" (IFSW, 2000). I am reminded of a verse from Leonard Cohen's *Anthem*: "There is a crack in everything. That's how the light gets in" (Cohen, 1993, p. 373).

On both the IFSW and IASSW websites, four paragraphs of commentary accompany the international definition of social work (IASSW, 2001; IFSW, 2000). For the most part, these elaborations merely re-enforce the social system focus of our profession's value base, foundation knowledge, and practice methods. But, once again, there are small openings to be found that could lead to broader interpretations of our role in the physical environment. First, the "s" is maintained as mention is made of "the multiple, complex transactions between people and their environments." Second, the theory base is said to include "local and Indigenous knowledge specific to its contexts." Third, our methods are portrayed as consistent with a "holistic focus on persons and their environments." More cracks!

While I am not blinded by the light, I can at least see the welcome rays creeping in. Although the physical environment is still not mentioned specifically, can it be far away when we are speaking of holistic, complex transactions with our multiple environments that may be understood through contextual knowledge? Why did the IFSW not just include explicit reference to the physical environment in its international definition of social work? This may be another example of the disfluency to be discussed in Chapter 8. Faced with the risk and anxiety of new situations, we fall back on our familiar patterns and easy words rather than fully expressing what we want to say.

Fuller expression is certainly evident in a later document from the IFSW, the *International Policy Statement on Globalisation and the Environment* (IFSW, 2004). If the IFSW's international definition of social work presented a few cracks through which glimpses of the physical environment might be possible, then the *International Policy Statement on Globalisation and the Environment* is a very bright light. It begins with a clear understanding that human rights and individual growth cannot be realized without supportive circumstances that include "confidence in a sustainable natural environment which supports life" (paragraph 18). The IFSW recognizes "that the natural and built environments have a direct impact on people's opportunities to develop and achieve their potential, that the earth's resources should be shared in a sustainable way" (paragraph 9). Individual social workers and their organizations are called on to recognize

> the importance of the natural and built environment to the social environment, to develop environmental responsibility and care for the environment in social work practice and management today and for future generations, to work with other professionals to increase our knowledge and with community groups to develop advocacy skills and strategies to work towards a healthier environment and to ensure that environmental issues gain increased presence in social work education. (paragraph 15)

The policy statement goes on to acknowledge that early social work did in fact concern itself with housing and the physical environment, but that this focus was dropped in the mid 20th century due to a preoccupation with the social environment. A case is made for returning to our roots and calling attention back to the physical environment as a legitimate and necessary focus for social work:

> Our communities have been rediscovering that a positive social environment is not possible without a sustainable natural environment. It is generally accepted that our natural environment not only influences but also is crucial for our social lives now and in the future. (paragraph 27)

These passages from IFSW's *International Policy Statement on Globalisation and the Environment* (IFSW, 2004) appear to confirm the centrality of the physical environment for social work's mission in the 21st century. How curious then that the concluding summary of the document makes absolutely no mention of the physical environment:

> This paper has presented a simplified analysis of dominant economic theories and strategies over the last fifty years and highlighted some of the adverse impacts of current policies of neo-liberalism/structural adjustment. It has identified the potential for change at international, national, and local levels. It has set out the contribution which social work can make through work with individuals, social development and community action in enhancing life choices for socially excluded individuals. (paragraph 70)
>
> IFSW will play its part in building a global coalition to promote an inclusive community. (paragraph 71)

What happened here? Are we back to smoke and mirrors? Has international social work brought the physical environment to centre-stage only to make it vanish once again? Are we witnessing disfluency—a retreat to our easy words and familiar perspectives in a risky situation? Given the eloquence and power with which the natural world is acknowledged, respected, and incorporated in the body of the policy statement, I do not think that the flawed summary will have a large negative impact. The light is already shining through. Yet I find it both disappointing and frustrating that the document summary is limited in this way. It will be more difficult for our international voice to lead us to an effective role in environmental concerns if we find the same trickery and disfluency at the international level that we have seen so clearly in the literature of the mainstream profession.

Although the mainstream social work literature has largely ignored the physical environment, we have seen some evidence that a sense of place and its importance for practice is beginning. Rural/remote social work has elaborated this sense of place in terms of context, stewardship, and responsibility. Deep ecology and Aboriginal social work have asserted the oneness of humans and the natural world. International social work has confirmed the importance of a sustainable natural environment and a healthy built environment. These are all encouraging developments, but the overall profession has not yet woven these strands together into a coherent whole, an understanding of people and place. The next chapter explores how other related disciplines are expressing and working with a sense of place.

CHAPTER SEVEN

PERSPECTIVES FROM OTHER DISCIPLINES

The Environment and a Sense of Place

Separations of the social from the natural,
biological and scientific can be effective in terms of
establishing disciplinary boundaries and gaining
professional recognition from practitioners of
other disciplines.... However, such division of
labour becomes profoundly problematic when
confronting environmental questions.

(IRWIN, 2001, P. 7)

n the first three chapters of this book, we saw how mainstream social work has generally neglected the physical environment, although there were a few calls to seek new and expanded ecological models that accept the interrelatedness of social and environmental issues. The next three chapters brought forward some intriguing concepts from the margins of our profession, including such notions as locality, context, place identity, place attachment, stewardship, spiritual landscapes, sacred sites, geopsyche, environmental citizenship, and sustainability. All of these concepts appear to have some connection with a sense of place, but that notion has not been well developed in the social work literature.

Perhaps it would be helpful at this point to pause and look outside of social work. How have other disciplines and professions responded to environmental concerns, and how have they expressed this sense of place in their own models and experiences? To explore these questions, I begin with the arts. At first glance, endeavours such as painting, film, music, and viticulture may appear to be very distant from our applied discipline of social work, but I have discovered potentially useful insights from these activities. I then move on to look briefly at two familiar disciplines that have given much to the knowledge base of social work—sociology and psychology. Finally, I turn to the related applied disciplines of planning, geography, and education.

Obviously this list of disciplines is not exhaustive, and I have not attempted to present a comprehensive literature review within each. Instead, I have tried to capture some of their understandings of the physical environment and sense of place that might be useful for social work as we approach our own model-building tasks ahead. Rather than arguments designed to prove something about place, these are snippets intended to stimulate thinking about place and placemaking processes.

PAINTING: ENCOUNTERS WITH LANDSCAPE

> The environment exists because it was made
> visible by the act of making it separate.
> (EVERNDEN, 1985, P. 126)

Reading in art history, I discovered that the Western assumption of a split between person and environment can be traced back to the Renaissance. Evernden (1985) examined Leonardo da Vinci's painting *Mona Lisa* and wondered if this may have been the point where the individual and the landscape emerged as separate entities:

> The famous enigmatic smile reveals a realm of privacy which we can glimpse but never know or possess, and the true individual is born.

But the individual is created by pulling significance inward, and nature retreats outward as the thing we know as landscape. (p. 126)

Note here that Evernden was not suggesting that da Vinci created the separation himself, but rather that his work reflects a transformation that was taking place in his society. Van den Berg (1961) had a similar reaction to *Mona Lisa*:

> The landscape behind her is justly famous; it is the first landscape painted as a landscape, not just a backdrop for human actions: nature, nature as the middle ages did not know it, an exterior nature closed within itself and self-sufficient, an exterior from which the human element has, in principle, been removed entirely. (p. 231)

[handwritten margin note: 2..? historical context]

It would seem that the separation of the person from the environment that I have observed in the mainstream social work literature has actually been a long-standing cultural assumption in the Western world. In social work, however, we have followed the separation by concentrating on the full and self-sufficient person, while virtually ignoring the environment from which the person has been conceptually removed.

Lee (2004) examined the tradition of Dutch landscape painting and argued that the landscape must be more than a distinct abstraction separate from the people who perceive it. Landscape is "a way of *thinking* about the land," a set of "creations of memory, dreaming or ritual" whose meanings are "in the eye of the beholder" (p. 404). This notion of landscape expresses placemaking or the assignment of meaning to space, rather than simply painting still landforms. I see such subjective expression of encounters with a landscape (rather than simply painting a picture of a scene) as parallel to Morito's (2002) activity of thinking ecologically (as opposed to simply thinking about ecology) that was presented in the Introduction to this book. Expressing a place appears to be a process, a relationship, rather than mere production of a product.

FILM: TELLING A PLACE

"Ya know, I think if ya live someplace long
enough, y'are dat place." —Rocky Balboa
(WINKLER, 2006)

Not widely recognized as a phenomenological philosopher, fictitious boxer Rocky Balboa effectively expressed a fundamental truth when explaining to his brother-in-law Paulie why he continues to live in the Philadelphia projects after so many years of fame and opportunity. This profound connection with place echoes the place-identity language of the Aboriginal authors discussed in Chapter 5.

Noted German film director Wim Wenders distinguished between "telling a place" and "telling a story" (p. 6) in a talk on "A Sense of Place" given at Princeton University in 2001. In his work, Wenders is drawn to places. When he finds a place that intrigues him, he senses that he will be the instrument for telling the place. Only after he comes to some understanding of the place does he look for characters. Eventually, after place and character, he may develop a story to tie everything together, but often he films his characters interacting in place without a script. Wenders contrasts his "telling a place" approach to the mainstream American cinema, where the story has become the driving force and places are relegated to scenery rather than the source of the story. Places have been "turned from 'instigators' to 'background'" (p. 14). According to Wenders (2001), the globalized commercial audiovisual industry has

> established "The Story" as the paramount force to move imagery and imagination, at the expense of the story-building power of people and landscapes. That shift will drastically shape and form future generations. Not only their imagination, but ultimately their image of themselves, their self-respect, and their knowledge of our common place, planet Earth. (p. 15)

Beginning with an understanding of place, then adding characters and possibly a narrative direction—this reminds me of the accounts of rural and Aboriginal social work practice in the literature. The worker starts with an understanding of the locality and the people, a living embeddedness. Then some incident or situation (the beginning of the narrative) arises that calls for a professional helping response. Help may be directed towards an individual or family (and a file may be opened in their name), but this specific work on "The Story" is connected to the context or place that existed prior to the individual story and will exist long after. The place is not "background." It is an "instigator" of the situation and the response.

In mainstream urban social work practice, the situation is often reversed. Contact between worker and client is instigated by the narrative, some situation that brings them together for the first time. An investigation or assessment may involve the worker learning about the client's environment, but this is background information and likely limited to aspects of the social environment. There is little expectation or recognition of any shared locality. Wenders' notion of telling a place rather than telling a story runs counter to the dominant form of discourse in mainstream social work.

MUSIC: SOUNDSCAPES AND EXPRESSIONS OF LOCALITY

> Both as a creative practice and as a form of
> consumption, music plays an important role in the
> narrativization of place, that is, in the way in
> which people define their relationship to
> local, everyday surroundings.
> (WHITELEY, 2004, P. 2)

Music is consumed daily in the experiential places where people define their relationships to their surroundings. We are all familiar with music as an expression of national identity through national anthems, sports allegiances through team songs, and product loyalty through advertising jingles. There is growing interest in the role of music in the understanding and expression of a person's relationship with locality, an intersection of music and place. Lewis (1992) concluded that "people look to specific musics as symbolic anchors in regions, as signs of community, belonging, and a shared past" (p. 174). A compilation from Warner Music Canada (2004), entitled *A Sense of Place: Music from and Inspired by the People and Places of Canada*, is a diverse collection of the musical expression of such regional identities.

Webb (2004) wrote of "the importance of place, location and identification with different elements of popular music genres and their influence on the sound of a locality's popular music" (p. 66). Can you imagine the blues without the Mississippi Delta? Or Dixieland jazz without New Orleans? The Beatles without Liverpool? Ladysmith Black Mambazo without South Africa? Certainly these musicians have performed and sold music around the globe, but the music itself was created in specific places within specific located circumstances.

"Community music" is gaining recognition as a field of research and practice (Higgins, 2002), but I have found no clear definition of the phenomenon. It appears to refer to music creation, performance, and education that are connected with particular settings, with place. Community music may be linked to performances at local celebrations or events. It may be created to promote local identity either through the content of the music itself or the interaction of local musicians. Teaching of community music can involve consideration of local history, spirituality, and patterns of daily life, along with the more conventional music theory, appreciation, and skills.

According to Whiteley (2004), music plays a "significant part in the way that individuals author space, musical texts being creatively combined with local knowledges and sensibilities in ways that tell particular stories about the local, and impose collectively defined meanings and significance on space" (p. 3). Drawing on his experiences as a researcher and performer with the three-stringed *lyra* on the Greek island of Crete,

Dawe (2004) connected lyra music with an "all pervasive and largely village-based moral geography" (p. 56) that defines the island. Sonic environments or "soundscapes" (Shelemay, 2001, p. xiii) can accommodate local detail while allowing for connections across localities. Explaining the importance of music to identity and patterns of everyday life, Finnegan (1989) introduced the notion of "musical pathways" to describe relationships between various musical expressions in any one locale. Musical pathways appear to be similar to the Aboriginal "songlines" (Chatwin, 1987) discussed in Chapter 5.

Mainstream Western culture and technology allow for increasingly efficient separation of musical expression from the geographic context in which the music was created. Through music downloads and satellite radio transmissions, many people can create and surround themselves with their own personal life soundtracks divorced from the immediacy of the local environment. One can now listen to urban rap or reggae while in the Yukon wilderness. This is not necessarily a bad development, but it is one more instance of us escaping or avoiding our local connections and interrelationships. Finding reference to authored space, musical pathways, and moral geographies in the music literature brings us back to a fundamental sense of place and identity for which music can be both an expression and a contributor.

VITICULTURE: *TERROIR* AND PLACE VALUE

> Divorced from its geographic origins, wine is only
> marginally more interesting than fruit juice, lager or gin.
> (GOODE, 2002, P. 1)

While working on this book, I was coincidentally introduced to the French concept of *terroir*, a term used among wine aficionados to suggest the possibility of capturing place value within a bottle. It seems that, within viticulture (the study of growing grapes), some have made a distinction between two genres of wine: commodity wines and *terroir* wines. Commodity wines are purchased because they fill a need or serve a purpose, much like any other food commodity purchased at the grocery store. *Terroir* wines, on the other hand, are purchased for their intrinsic qualities derived from factors such as the type of grape, vineyard location, soil type, altitude, drainage, pruning philosophy, local microclimate, and sun exposure. According to McGee and Patterson (2007), *terroir* refers to "the relationship between a wine and the place it comes from" (p. 76). In this way, *terroir* wine is linked to geography, to specifics of the environment.

The contents of a sealed bottle of *terroir* wine express the geographic reality of a particular place and time. *Terroir* can be understood as "the possession by a wine of a sense of place, or 'somewhereness.' That is, a wine from a particular patch of ground expresses characteristics related to the physical environment in which the grapes are grown" (Goode, 2003, p. 1). The dominant perspective in industrial society regarding viticulture had been that grape growing is the simple production of raw material to be sent to the specialist winemaker. *Terroir* is a very different understanding. *Terroir* wines have place value, a geographic basis whereby each bottle captures a unique environmental reality.

Debate continues about the definition of *terroir*. Some view it as a term encompassing all of the environmental conditions that contribute to the production of a bottle of wine; others include the human techniques applied in the process. A retrospective by Bohmrich (2006) referred to the difficult problem of "how to separate the influences of nature and nurture" (p. 2). A *terroir* hierarchy was proposed, ranging from macro-*terroir* (country, region) to meso-*terroir* (district, locality) and micro-*terroir* (site). A *terroir* model (p. 6, 7) was proposed to account for the relative contributions of natural conditions and human intervention. The model identifies three categories of wine: (1) highest-quality wine of specific and identifiable character; (2) average commercial-quality wine with varietal and/or regional identity; and (3) base generic wine.

I am quite familiar with the nature/nurture debate in the social sciences, but I was surprised to discover the same issues in viticulture! Viticulture is engaged in not only a familiar debate about the impact of natural vs. human influence, but also in an application model using the micro–meso–macro perspective that is at the core of social work. These parallels make me wonder about possible applications of the *terroir* model in social work. Could it be that our conventional person-in-environment perspective, with its limited focus on social environments, leads us to consider clients as "base generic" interchangeable people, economic units requiring a generic fit with society? Maybe our recent attention to diversity issues allows us to roughly appreciate "varietal and/or regional identity" and begin to consider context. But we seem to be a long way from considering the "specific and identifiable character" of a client based on the places he or she has come from and the places where he or she makes a life.

SOCIOLOGY: *HABITUS* AND EMBODIED SENSE OF PLACE

> The ambitious argument for an environmental
> sociology suggests the need to transcend rather
> than be constrained by established intellectual and
> disciplinary boundaries—even if ... this has significant
> implications for the discipline as a whole.
> (IRWIN, 2001, P. 7)

In a major address to the British Sociological Association, Newby (1991) claimed that his discipline's contributions to the environmental threats facing the world had been "disappointing" (p. 1). He blamed sociology's silence on fundamental theoretical assumptions that had to be challenged and overcome. Newby called for sociology to "lose no opportunity to acquire the appropriate knowledge about ourselves and our relationship to the planet" (p. 8). Several volumes followed, including *Social Theory and the Global Environment* (Redclift & Benton, 1994), *An Invitation to*

Environmental Sociology (Bell, 1998), and *Sociology and the Environment* (Irwin, 2001).

These books explored the social construction of nature and environmental issues. In addition, they considered matters of reconstructing theory within the discipline of sociology itself. Very similar to social work, sociology acknowledged its long-standing focus on the social environment to the exclusion of the physical environment, which was regarded as "asocial and external to human life" (Irwin, 2001, p. 3). There was a new goal:

> [To] find another relationship to nature besides reification, possession, appropriation and nostalgia. No longer able to sustain the fiction of being either subjects or objects, all the partners in the potent conversations that constitute nature must find a new ground for making meanings together. (Haraway, 1995, p. 70)

Sociology has also contributed to the environmental discussion through the concept of *habitus*, defined in the language of academic sociology by Bourdieu (1990) as "a system of durable, transposable dispositions, structured structures predisposed to function as structuring structures, that is, as principles which generate and organize practices and representations" (p. 53). Fortunately for those of us who do not speak sociologese, Hillier and Rooksby (2005) further translated Bourdieu's *habitus* as "a sense of one's (and others') place and role in the world of one's lived environment ... an embodied, as well as a cognitive, sense of place" (p. 21). *Habitus* thus conveys a sense of the connection between person and place, not only in terms of identity, but also with attention to the life opportunities in a particular local environment.

With the exceptions of Coates' (2003) groundbreaking book and a few contributions that were discussed in Chapter 3, the social work literature of the last decade has presented very little challenge to its long-standing silence on environmental issues facing the planet. We have not yet produced volumes of readings on environmental social work similar to the activity in sociology. Might we come to an acceptance of the located and embodied person in relation to a living environment? In social work, we have only begun to explore a sense of place, and we might do well to look once again to sociology and its understandings of *habitus*.

PSYCHOLOGY: ENVIRONMENTS, PLACE, AND BEHAVIOUR

> A conceptual topic of continuing interest within environmental psychology is the concept of place. How are places developed, how do they acquire meaning to people, how are they related to people's plans of action, their preferences, and even to their emotional reactions and well being?
>
> (EVANS, 1996, P. 4)

Psychology and social work are often considered as sister professions that share similar knowledge bases and practice skills. According to Heft (2001), the Western separation of person from environment can also be found at the core of contemporary psychology. This division may hinder psychology from perceiving a meaningful natural world, an environment of "meaningful places" (p. 329). Environmental issues identified in psychology appear, at first glance, to be very similar to those in social work. Yet psychology seems to have made a deliberate effort to incorporate and address these concerns through a declared specialization of environmental psychology.

Just what is environmental psychology? One of the first clarifications explained that environmental psychologists were "concerned with human problems in relation to an environment of which man is both victim and conqueror" (Proshansky, Ittelson, & Rivlin, 1970, p. 5). I take this to mean that humans could be either the subject or the object of active encounters with the natural world. Either can be the instigator or the background in their mutual interactions. A later definition appeared as an entry in the 1999 edition of the *Encyclopedia of Environmental Science*: "Environmental psychology examines the interrelationship between environments and human behavior. The field defines the term environment very broadly including all that is natural on the planet as well as social settings, built environments, learning environments, and informational environments" (DeYoung, 1999). The American Psychological Association *Dictionary of Psychology* (VandenBos, 2007) defined environmental psychology as "a branch of psychology that emphasizes the effects of the physical environment on human behaviour and welfare" (p. 335). Also included was an entry on "proecological behaviour," defined as "behaviour that promotes the quality of the natural environment" (p. 737). I note that over time these definitions of environmental psychology have moved from a "concern" to a "field" to a "branch" of the profession, indicating that the specialized focus has become more known and accepted.

The profession has also developed several journals charged with building a relevant knowledge base for environmental psychology. *Environment and Behavior* is "designed to report rigorous experimental and theoretical work focusing on the influence of the physical environment on human behavior at the individual, group, and institutional levels" (Bechtel, 2008). The *Journal of Environmental Psychology* is devoted to "the study of the transactions and interrelationships between people and their sociophysical surroundings (including planned and natural environments) and the relation of this field to other social and biological sciences and to the environmental professions" (Gifford, 2008). *Ecological Psychology* pursues "the understanding of psychological and behavioral processes as they occur within the ecological constraints of animal-environment systems" (Mace, 2008). An entire special issue of *American Behavioral Scientist* (Tigges, 2006) was devoted to a sense of place and community cohesion. In addition to these journals, a directory of credit course offerings linking psychology and environmental issues addresses such areas of study as conservation psychology, ecological psychology, environmental psychology, and the psychology of sustainability and behaviour change (Scott & Kroger, 2006).

It could be argued that psychology has effectively relegated environmental issues to the margins of the profession by creating and labelling the environmental psychology specialization, much as we have done in social work with our specializations of rural or Aboriginal social work. In this way, environmental issues might be contained at the boundaries of the profession with little chance of them contaminating the mainstream urban clinical thrust. It appears to me, however, that the efforts in psychology deserve more credit. The journals and courses devoted to environmental concerns support the process of creating a knowledge base with the potential to have a significant impact on the mainstream profession and beyond. The profession of psychology seems to clearly understand that its environmental activities must be multidisciplinary in nature. Environmental psychology "must be understood in the context of the overall environmental sciences ... [part of] the large body of study concerned with the consequences of man's manipulation of his environment" (Bonnes & Bonaiuto, 2002, p. 28). In social work, we might do well to look to environmental psychology for content on places and meaning, for perspective on the multidisciplinary nature of environmental endeavours, and for strategies for systematically building our knowledge base.

ENVIRONMENTAL DESIGN: PLACEMAKING AND POSITIVE SPACE

> Designing and building healthy places is not a new concept; for centuries, those who care about health, across the professions, have turned their attention to the built environment. We are now rediscovering some of this old wisdom, and identifying principles for healthy placemaking for the new century.
> (FRUMKIN, FRANK, & JACKSON, 2004)

The history of environmental design, in particular the planning profession, can be traced back to the industrial revolution (p. Hall, 1996) when displaced workers and families warehoused in slum conditions bred poverty, crime, and addiction. It was thought by some that the impact and experience of these social ills might be lessened through a process of *placemaking*, in particular the thoughtful planning of parks and gardens in urban areas to create healthier local communities. This strong sense of the importance of place and the responsibility to create spaces with positive meaning has remained at the core of the planning profession. Gauldie (1969) spoke of how life "in an environment which has to be endured or ignored rather than enjoyed is to be diminished as a human being." Architect Norberg-Shulz (1980) offered a very concise definition of place as "space plus character" (p. 18). It would appear that the concept of place here involves both location and ascribed meaning. More poetically, Leach (2005) reflected on how "memories of associated activities haunt architecture like a ghost" (p. 308).

Bartuska (1994) challenged designers and planners to "design and manage with people and with the built and natural environments. If we succeed, we will be able to celebrate sustainable life throughout the biosphere. If we fail, it may lead to our demise" (p. 170). These challenges faced by planners and architects, challenges to integrate the physical and social, are similar to those we have seen in other disciplines. Their profession's goals are social, but its methods are physical. For a long time, the complexity of these challenges led many planners to perceive the social context as outside the scope of their profession (Taylor, 1999). That is changing. As Brassard (2002) observed:

> Environments that are created without acknowledging individuals as the "public" and therefore as the primary stakeholder, do not sufficiently support individual development towards self-realization and fulfillment. Most of our physical communities are the result of a select few individuals, usually professionals or experts, evaluating this expression and making decisions as to what that expression means. (p. 54)

Clark and Stein (2003) confirmed the necessity for public managers to consider the emotional connections of local stakeholders to their specific cultural landscapes. They presented their argument using terms such as "sense of place," "community attachment," and "place attachment."

Other recent developments also challenge any assumed division between human and physical environments. With the provocative title of *Design Like You Give A Damn: Architectural Responses to Humanitarian Crises*, Architecture for Humanity (2006) presented a strong case for humanitarian architecture and socially conscious design. The authors argued that the physical design of the built environment shapes how we live our lives, but observed that such architectural and planning resources are not affordable in many of the regions where they are most desperately needed. Nearly 50% of the world's population has no access to clean water or effective sanitation, while some 15% live in refugee camps or slums (Architecture for Humanity, 2006, p. 1). The book showcases innovative international projects that have promoted the meeting of basic human needs and sustainability. Even in this work on sustainability, place remains a key concept. This echoes Barton's (2000) framework for planning sustainable communities and "eco-neighbourhoods," which advocated a hub approach beginning with a spatially defined central core where interactions lead to identity and the meeting of needs.

Planners and social workers have much in common. Both professions are concerned with persons and environments. Social work struggles to incorporate the physical environment; planning struggles to incorporate the human element. In both professions, broad visions for a better world are often constrained by the practical realities of public policy, budgets, and bureaucracies. In recognition of these shared difficulties, a new course called "People and Place" at the University of Calgary, co-taught by professors from the Faculty of Social Work and the Faculty of Environmental Design, brings together graduate students from both disciplines to explore the human dimensions of the built environment as well as the environmental context of human interaction (Zapf & Rogers, 2006).

GEOGRAPHY: EARTHKEEPING AND PLACES THAT MATTER

Humans, as members of groups, create places; in turn,
each place created develops a character that affects human
behaviour. This circularity is characteristic of all landscapes,
but it is perhaps most relevant in the extreme cases—
landscapes of the advantaged and the disadvantaged,
the privileged and the underprivileged, insiders and
outsiders, rich and poor, and men and women.

(NORTON, 2004, P. 264)

Looking back, I believe that high-school geography left me in a state of "geographic illiteracy" (de Blij & Murphy, 2003, p. 6) with no real understanding of human settlement patterns or regional interactions. My recollections of academic geography involve long lists to be committed to memory (capital cities, annual precipitation, landforms, economic products) and intimidating testing situations where we had to place labels on blank maps. All of this contributed to "the popular stereotype that geography is a discipline steadfastly devoted to long factual inventories and rote memorization" (Taaffe, Gauthier, & O'Kelly, 1996, p. 3). Geography was also presented to me in school as a stand-alone subject with no connections or relevance to the other subjects I was learning. I do recall some minor efforts to combine history and geography under the general heading of "social studies," but my high-school report cards included a grade assigned for geography alone so naturally it took on the status of a distinct discipline in my mind.

I also vividly recall the coloured map of Canada that was featured prominently on the wall of every grade-school classroom I attended. This map was presented as the reality of my country, a reality that I accepted without question. There was Canada, colourful but with no neighbours on three sides. To our north, east, and west, all I could see was a thin strip of ocean giving way to the nothingness of the bare wall supporting the map. It was only to the south that I saw a neighbour. My country was firmly anchored to the mainland United States. No one told me then that this was only one of many possible perspectives. I was told this was Canada. After many years as a northern social work practitioner and educator, I can now see that grade-school map for what it was. Looking, for example, at a different projection with the North Pole at the centre, I can easily see our neighbours in Siberia, Scandinavia, Greenland, Iceland, and Alaska, and I become more open to perceiving the common issues we face related to the Arctic Ocean, climate change, pollution, development, and resource management. I can also see now how the convention of presenting the provinces and territories in different colours inhibited my ability to perceive the regional nature of Canada and the core/peripheral relationships that have shaped us (Bone, 2002).

Modern geography has been described as both a natural/physical science and a social science (de Blij & Muller, 1994; Rubenstein, 2005). The discipline's interests, "while seriously preoccupied with the Earth's environment, are also closely related to the mainstream concerns of the social sciences" (Mabogunje, 1996, p. 447). Geography is "about how, why, and where human and natural activities occur and how these activities are interconnected" (Strahler & Strahler, 2005, p. 6). Current geographic frameworks focus not only on physical geography (e.g., landforms, coastlines, climates, soils, vegetation, animals) but also on human geography (e.g., culture, language, development, spatial patterns of human activities, meanings attributed to places). The distinction between physical and human geography, however, appears not to be a total separation but more a matter of different emphasis (Norton, 2004; Rubenstein, 2005).

I find evidence in the geography literature that the concept of "place" is a major point of intersection between physical and human geography. In the mid 1980s, a project involving the National Geographic Society proposed a set of five themes that were considered fundamental to the discipline of geography (Geography Education National Implementation Project, 1986). Two of these themes were *human–environment interactions* (mutual influence of human groupings and the natural environment) and *place* (meaning of particular locations distinguished by human activity and physical characteristics). Another group working in the 1990s identified three spatial themes that were said to integrate human and physical geography (National Research Council, 1997): *integration in place* (mutual influence of people and things located in the same place, interactions to develop the character of a place); *interdependencies between places* (networks and interactions that connect different places); and *interdependence among scales* (local, regional, and global circumstances influencing the character of place). Norton (2004) affirmed that "space and place are constantly being reinforced as our key concepts" (p. 509).

Just what is this notion of *place* in the geographic sense, and how does it differ from related concepts such as space, location, and site? Geographers tell us that every point on the surface of the Earth is unique and can be described in terms of location (a mathematical representation using meridians and parallels), site (physical characteristics, landforms), situation (relative to other points), space (physical intervals or gaps between points), and maybe even a name (Rubenstein, 2005). Essentially, these descriptors identify aspects of physical location. What then is place? Rubenstein (2005) clarified that "to geographers, a *place* is a specific point on Earth distinguished by a particular character" (p. 5).

It would appear that place has something to do with location plus the meaning ascribed to the location. This notion of place is supported by other geographers (as well as by the environmental designers in the previous discussion who described place as space plus character). Beyond the simple location of objects, place "is also a reality to be clarified and understood from the perspectives of the people who have given it meaning" (Tuan, 1974, p. 213). To Entrikin (1997), "the geographical concept of place refers to the areal context of events, objects and actions. It is a context that includes natural elements and human constructions, both material and ideal" (p. 299). Agnew and Duncan (1989) presented the notion of place as an integration of "the geographical and sociological imaginations." Norton (2004) considered place as location plus "the values that we associate with that location," or "a location that has a particular identity" (p. 56).

Placemaking could be more complex than it first appears. Assigning meaning to particular locations may not be a uniform or consensual process; different groups can ascribe different meanings or significance to the same region. Their experiences and history with the region may be very different. Some groups have more power than others and may be able to impose their meanings. As Lee (2004) observed, the process of assigning meaning to space, the process of making places, involves "the layering of values and the potential for conflicting values, even between members of the same cultural group" (p. 404).

The geography literature also identifies a related notion of *sense of place,* or "the attachment that we have to locations with personal significance" (Norton, 2004, p. 56). According to Rubenstein (2005), "humans possess a strong sense of place—that is, a feeling for the features that contribute to the distinctiveness of a particular spot on Earth" (p. 15). Sense of place then appears to be a phenomenological concept, a consideration of the meanings that places hold for people, an acknowledgement of identification with place. As people perceive and subjectively experience their physical environments, places will assume meaning for them. From the perspective of phenomenological geography, the essence of place can be found

> in the experience of an "inside" that is distinct from an "outside"; more than anything else this is what sets places apart in space and defines a particular system of physical features, activities, and meaning. To be inside a place is to belong to it and to identify with it, and the more profoundly inside you are the stronger is the identity with the place. (Relph, 1976, p. 49)

Examining the impact of the cod-fishing moratorium in Newfoundland and Labrador, Canning and Strong (2002) presented a powerful picture of becoming an outsider in one's home. Paying the human costs of unemployment and out-migration, local fishers were separating from their social supports, life patterns, and geographic sense of identity. They were becoming "disembedded" (p. 323). As the local meanings were withdrawn, they lost their sense of belonging and sense of place.

Trying to develop new language for such discussions of place, an imaginative book on personal geographies (Harmon, 2004) introduced concepts of body maps, mental geographies, maps of the imagination, and mental maps. Conventional pursuits such as navigation, psychology, and metaphysics were creatively combined in a new concept of "orientating" (Hall, 2004), which involved

> crashing through the larger landscapes of memory and experience and knowledge, trying to get a fix on where we are in a multitude of landscapes that together compose the grander scheme of things. Orientating begins with geography, but it reflects a need of the conscious, self-aware organism for a kind of transcendent orientation that asks not just where am I, but where do I fit in this landscape? Where have I been? Where shall I go, and what values shall I pack for the trip? What culture of knowledge allows me to know what I know, which is often another way of knowing where I am? And what pattern, what grid of wisdom, can I impose on my accumulated, idiosyncratic geographies? The coordinates marking this territory are unique to each individual and lend themselves to a very private kind of cartography. (p. 15)

Some geographers have attempted to apply the conceptualization of place to other disciplines where "the role of experiences and representation of place remains undertheorized" (Bauder, 2001, p. 37). Others have added a component of time to this sense of place in order to incorporate heritage, or the development of a place identity over time (Ashworth &

Graham, 2005). Still others have added a notion of moral responsibility in the form of active "stewardship." Expressed by Lerner (1993) as "active earthkeeping," stewardship appears to feature active local expression and a specific connection with place in addition to an overall global responsibility for the planet. "Stewardship permits people to take leadership roles and act responsibly when a threat to a locally valued place or environment occurs" (Draper & Reed, 2005, p. 25).

In order to achieve a goal of sustainability, humans must undergo a transformation from "managing resources to managing ourselves" while we "learn to live as part of nature" (Wackernagel & Rees, 1996, p. 4). Given our history of separation between person and nature, and the momentum of our established mindsets, social structures, and physical structures that support the split, this ideal of living as part of nature presents a special challenge in the Western world. Draper and Reed (2005) offered a direction:

> Humans could learn to live sustainably if they understood and mimicked how nature perpetuates itself. Learning to live sustainably begins with recognizing the following: humans are a part of, and not separate from, the dynamic web of life on Earth; human economies, lifestyles, and ultimate survival depend totally on the sun and the Earth; and everything is connected to everything else, although some connections are stronger and more important than others. (p. 91)

I see many parallels between geography and social work. Both are concerned with the interactions between people and the environment at multiple levels (micro, mezzo, and macro in social work; local, regional, and global in geography). Both social work and human geography draw on similar interdisciplinary foundations for their knowledge bases (anthropology, economics, sociology, psychology, biology, political science), and yet seldom acknowledge each other in theory and practice writings. An ecosystems framework has been influential in both disciplines, and it appears both are coming to understand the importance of people and place as an organizing concept.

Borrowing language from British geographer Stephen Pile (1997), I have written elsewhere about how geography has placed an emphasis on *peopled places* while social work has focused on *placed peoples* (Zapf, 2005b). De Blij and Murphy (2003) suggested that geography may be looked to for new ideas and approaches as the world comes to terms with environmental crises, because geography has recognized that "place matters, as does the relationship between phenomena in place and space" (p. 49). It appears to me that integration of the natural and the human has been more comprehensive in geography than in social work. Perhaps their language of *place* has been more open than our limited notion of *environment*. That may be why issues of sustainability and stewardship seem to be more visible at the core of geography, rather than at the margins as in social work.

EDUCATION: WAYFINDING AND LIVING WELL IN PLACE

All education is environmental education. By what
is included or excluded we teach the young that they
are a part of, or apart from, the natural world.
(ORR, 2004, P. 1)

Much of the education literature focuses on curriculum content and learning processes, but there has been some interest in the physical context of education. Issues of school architecture and classroom design are frequently discussed. Perkins (2001), for example, examined various physical forms that schools might take and how these forms conveyed meaning for the students individually and for the larger education enterprise. Perkins offered a process of *wayfinding* to describe how people find their way through a place by negotiating various pathways, boundaries, and signs of meaningful activity.

I find an emphasis here on the notion of schools as educational places, built spaces that have meaning for individual students, the community, and the education system. Hutchison (2004) explained how place can be individually constructed, "a reality informed by the unique experiences, histories, motives, and goals that each of us brings to the spaces with which we identify" (p. 11); place can also be socially constructed by groups of people who have come to a general agreement about the defining boundaries and functions of a particular space (e.g., school, mall, park). A particular space becomes a place as it develops an emotional significance, a meaning or spirit based on the activities carried out there. This perspective on creating places in education appears very similar to the processes we have seen in geography and environmental design.

School design attempts to promote a spirit of place in addition to supporting the nuts and bolts of the education process. Classroom design, the careful planning and preparation of the educational environment, has many facets, including furnishings, physical walls and structures, multipurpose spaces, open spaces, energy efficiency, accessibility, community-use policies, health and safety issues, and aesthetics. School design has been described as "the intersection of architectural style, educational philosophy, demographics, and budgetary realities over time" (Hutchison, 2004, p. 49), or more concisely "the intersection of philosophy and place" (p. 80). Once again, place emerges as the location where people and meanings come together.

paradox

While most attention in the context of education has focused on the built environment of schools, some interest has been expressed in the natural environment surrounding school buildings. A strategy called "school-ground naturalization" (Evergreen Foundation, 1994) encouraged reclaiming concrete playgrounds and adjacent spaces by reintroducing Indigenous plant species. Diers (2004) described a community garden project developed jointly by a school and a neighbourhood association in Seattle. Caring for these naturalized spaces can provide meaningful outdoor learning and environmental education experiences for both students and local citizens. As Hutchison (2004) observed, "such projects help foster a healthy appreciation of nature and enhance a sense of community in the surrounding neighborhood" (p. 102).

Smith (2007) described an emerging curriculum development approach called "place-based education," which "seeks to link classrooms even more tightly to their communities and regions.... Place-based education works to cultivate students' knowledge of the unique characteristics of their home communities and to engage them in meaningful and authentic work" (pp. 20–21). The underlying philosophy acknowledges that humankind faces major environmental challenges on a global level. Placed-based education focuses on localities, neighbourhoods, and communities, attempting to "offer experiences likely to stimulate a connection to people and place" (Smith, 2007, p. 21). The goal is to teach young people environmental stewardship and social responsibility by addressing local issues, a sort of global citizenship through meaningful local citizenship that bypasses nationalism and corporate interests. At the centre of this education trend is the notion of *people and place*, education that fosters an "allegiance to particular communities" (p. 21) as an alternative to training for mobility and careers in the global market economy.

Orr (1992) was one of the first to proclaim the value of learning to live well in one's place. This notion of "living well in place" is a very different concept from the idea of "welfare" to which my profession of social work has been committed. Welfare, or "faring well," refers more to a state of well-being and freedom from oppression regardless of context; it is a quality of the individual or family. Living well in place, on the other hand, involves the building of sustainable communities. To Morito (2002), living well involved living in tune with the rhythms of nature. Lane (2001) observed that "to see the same place in a hundred different ways is much harder (and infinitely more rewarding) than visiting a hundred different places and never seeing any of them" (p. 255).

Advocating the need to re-educate people for learning to live well in place, Orr (1992) made a key distinction between residents and inhabitants:

> A resident is a temporary occupant, putting down few roots and investing little, knowing little, and perhaps caring little for the immediate locale beyond its ability to gratify. As both a cause and effect of displacement, the resident lives in an indoor world of office building, shopping mall, automobile, apartment, and suburban house and watches as much as four hours of television each day. The inhabitant, in contrast, "dwells," as Illich puts it in an intimate, organic, and mutually nurturing relationship with a place. Good inhabitance is an art requiring detailed knowledge of a place, the capacity for observation, and a sense of care and rootedness. Residence requires cash and a map. A resident can reside almost anywhere that provides an income. Inhabitants bear the marks of their places whether rural or urban, in patterns of speech, through dress and behavior. Uprooted, they get homesick. (p. 130)

A multidisciplinary collection of essays on community and place from Yale University (Vitek & Jackson, 1996) supported this notion of living well in place, expanding on the idea of "dwelling" or placemaking rather than simply living somewhere (Tall, 1996). Others have extended the concept of inhabiting one's place to introduce the possibility of "re-inhabiting" the places where we already reside (Kemmis, 1992; Spretnak, 1991) through a process of "human homecoming" (Grange, 1977, p. 146). This process involves learning "to live attentively in place" (Spretnak, 1991, p. 82), "learning anew to-be-at-home in the region of our concern" (Grange, 1977, p. 146), and eventually becoming "people of place" (Cajete, 1994, p. 85). Such a process of reinhabiting reminds me of the words of the poet T.S. Eliot: "And the end of all our exploring/Will be to arrive where we started/And know the place for the first time" (p. 145).

Orr (1994) questioned the basic purpose of education, given the imminent environmental crises and threats to our planet. Conventional education prepares students for competition in a global economy that is devouring resources at a rate that cannot be sustained. Orr wanted education to consider how students, graduates, and human organizations impact the natural environment. It is now necessary for education to move beyond conventional schools to engage the wider society in a transformation to ecological literacy, "the active cultivation of ecological intelligence, imagination, and competence" (Orr, 1994, p. 1). From the perspective of Indigenous education, Cajete (2000) called for the same sort of transformation: "Western society must once again become nature-centred, if it is to make the kind of life-serving, ecologically sustainable transformations required in the next decades" (p. 266).

Jane Jacobs (2004), in her warnings of the dark age ahead, cautioned us that Western universities were in the business of "credentialing, not educating" (p. 44). A credential is portable; most are standardized or accredited by a central body, indicative of the same accomplishments regardless of venue. Jacobs (2004) argues that such credentials may have value for the individual graduates and their families, but they "are not a good investment for society" (p. 79). Reinforcing old and possibly obsolete paradigms can inhibit the creative and holistic thinking that is necessary to clean up the mess we have made through our credentialed planners and policy makers.

At an international conference on rural communities in Nanaimo in 2000, I first encountered a short book published by the Appalachia Educational Laboratory called *Place Value: An Educator's Guide to Good Literature on Rural Lifeways, Environments, and Purposes of Education* (Haas & Nachtigal, 1998). When I refer to this book as short, I may be guilty of understatement—apart from a 35-page annotated bibliography, the text of the book runs to only 30 pages! Yet in this brief guide for rural educators, I discovered a very meaningful framework for thinking about people and place. The book's strength lies in its innovative organization and not its exhaustive content.

A product of its time, Haas and Nachtigal's (1998) book was a reaction to education trends that were destroying rural communities by educating students for profit and success in the mobile global economy and mass culture. Through conventional education processes, rural students were being educated for success elsewhere, usually involving a move to a city. *Place Value* sought to reverse this trend by challenging rural educators to reconsider the very goals of the education process. In the preface, rural teachers were encouraged to focus on questions such as "What do students need to know about living well in their own communities?" and what are "ways in which teachers can relate their work and lives to the places where they live and help students do the same?" (p. vi).

Haas and Nachtigal (1998) expressed their position concisely when they declared that their book "offers a variety of perspectives on what we think it means to live well in a particular place" (p. vii). Living well in a particular place! Here was a profound notion conveyed by six simple and understandable words. "Living well" captures the sense of human welfare, but in an active mode. This is not simply an achieved sense of well-being. It is more than the good life, a materialistic end state to be achieved. Living well is an ongoing process, a goal, a human endeavour. Attempting to live well in place can lead to healthy people and healthy environments. Active participating citizens can respect and contribute to the health of their immediate built and natural environments.

The first essay in *Place Value* promotes a sense of place through consideration of "education for living well ecologically" (Haas & Nachtigal, 1998, p. 1). Such education actively involves students in seeking to understand their local surroundings, "learning the history, ecology, and social and physical infrastructure that surrounds them" (p. 3). Concurrent with this focus on the local place is a global awareness that we are "living part of the earth's life" (p. 2) and that "the survival of humanity depends on our understanding of how ecosystems organize themselves" (p. 4). Truths about these larger connections with the planet may be revealed through openness and thoughtful study of the local environment.

With this sense of place as a foundation, Haas and Nachtigal (1998) proceeded to explore education for living well politically (a sense of civic involvement), education for living well economically (a sense of worth), education for living well spiritually (a sense of connection), and education for living well in community (a sense of belonging). This profound concept of people living well in place appears to integrate the issues scattered throughout the social work literature, both at the margins and at our core. Issues such as a sense of place, environmental citizenship, sustainable economic activity, spirituality, identity, and belonging—these are all incorporated in a perspective of people living well in place.

SUMMARY: PLACE, SUSTAINABILITY, AND MULTIDISCIPLINARY EFFORTS

What have we learned from this brief encounter with other disciplines? What concepts or processes are related disciplines using to understand the physical environment? How do they see their contributions to the environmental crises facing humankind? What might we learn that could be helpful for social work?

 We have seen that the separation of people from background environments is not simply a quirk of modern social work. Such separation has been characteristic of Western traditions since the Renaissance. A self-sufficient natural world has been expressed through works of art for centuries. Yet painted landscapes are not simply photographic likenesses of external locations. They involve the artist's (and the viewer's) relationship with nature. A landscape picture reflects and evokes a way of thinking about the land, assigning and conveying meaning to geographic spaces.

Modern filmmaking generally focuses on story or narrative, with location relegated to the status of background. But there are alternatives, such as "telling a place," whereby the setting or place assumes the role of instigator rather than background. Common places and shared located realities can be a starting point for understanding people, not a conclusion. Music is another way in which art can author space and assign significance to places. Music creation, performance, and education can all be expressions of place. The idea of negotiating one's way through the physical environment using music is expressed through concepts such as soundscapes, musical pathways, and songlines. The geographic reality of a bottle of wine is captured in the notion of *terroir*. Wine can have potential "place value" or "somewhereness." Some wines (beyond base generic commodity wines) can have identifiable characters based on influences of nature and nurture in particular places.

Recognizing that multidisciplinary responses will be required to address environmental concerns, both sociology and psychology are attempting to understand processes by which disciplines come together to make meanings in places. Sociology's notion of *habitus* highlights one's place in relation to local life opportunities, a sense of one's role in the lived environment. Psychology is exploring how meaningful places in the natural and built environments (sociophysical surroundings) can influence human behaviour.

Environmental design and geography are also exploring connections between human and natural processes. They clarify the perspective of place as location plus meaning or character. Active "placemaking" involves an understanding of the natural environment and the real involvement of stakeholders in creating meaning through purposeful planning of the built environment. Circularity is acknowledged as human groups create places that in turn influence the behaviour of those who use the places. Human identity can be influenced by a sense of place, a feeling of being inside or outside, of being at home or away.

Education also concerns itself with this notion of place as space plus emotional significance from the activities carried out in that place.

Learning to be a full human being involves a process of "wayfinding," or learning to negotiate places. Place-based education focuses on learning to live well in place, learning to inhabit rather than reside.

Across the areas of study considered in this chapter, and particularly the applied disciplines, three common themes are apparent. The first is an acceptance of place as a foundation concept that integrates human activity with the physical environment. The second is a vision of sustainability achieved through processes such as stewardship, earthkeeping, and living well in place. The third is a belief that multidisciplinary responses are needed to take on the challenges of the environmental crises we have created.

Where does all of this leave social work? Our limited focus on the social environment has prevented us from developing a full understanding of place as central to our work. We are beginning to see discussions of stewardship and sustainability at the margins of our profession, but so far they have had little influence on our mainstream theory and models of practice. It may be difficult for us to engage in multidisciplinary efforts on environmental issues until we have a more developed understanding of the connections between people and place. Our very vocabulary and language patterns could be hindering our ability to engage fully in this process. The next chapter explores communication obstacles to meaningful discourse on environmental issues in social work.

CHAPTER EIGHT

LANGUAGE AND DISFLUENCY

Expressing the Environment in Social Work

Translation was never possible.
Instead there was only
conquest, the influx
of the language of hard nouns,
the language of metal,
the language of either/or,
the one language that has eaten all the others.
(MARGARET ATWOOD, 1995, *MARSH LANGUAGES*)

n exploring the place of the physical environment in the literature of social work and related disciplines, I have so far focused primarily on the content of the discussion. I have presented evidence from the literature about what has been said (and not said) regarding the physical environment. Now I want to consider the process of the discussion. How do we speak (or not speak) about the physical environment? Ideas in the literature are conveyed through written language—words and symbols arranged according to conventional rules of grammar and punctuation. What impact might the form of expression have on our ability to understand and deal with issues of the physical environment? In other words, are there constraints inherent in conventional social work approaches to speaking and writing about the physical environment?

Consider this example from the recent Canadian Environment Awards 2007 (Prefontaine, 2007). One of the recipients of a Community Award for Sustainable Living was quoted as declaring "My goal is to leave no stone unturned in the name of reducing our impact on Earth" (p. 35). Turning over all the stones may not be the best image with which to convey reducing our impact on the planet. I do not mean to belittle the achievements of this most worthy award recipient. Rather, I want to illustrate a common dilemma where our language itself can inhibit what we are trying to express about the environment.

Now I want to return to social work and our foundation metaphor expressed as "person-in-environment," which contains two nouns and a preposition, all separated by hyphens. In this chapter, I argue that unintended messages may be conveyed by the very grammatical construction of the phrase itself. First, I examine the punctuation, the language, and the parts of speech. Then I explore some possible limitations of the English language itself for this work. I conclude with a discussion of disfluency and further observations about our apparent difficulties in social work when attempting to speak about the physical environment and our interconnections.

VISUAL CUES: PUNCTUATION AND CAPITALIZATION

In an unlikely bestseller concerned entirely with punctuation, Truss (2003) devoted an entire chapter to the hyphen. It seems the poor hyphen has been under siege for some time. Truss quotes authorities ranging from Churchill's depiction of the hyphen as "a blemish, to be avoided wherever possible" (p. 168) to the 2003 edition of the *Oxford Dictionary of English*, which predicted that the hyphen is "heading for extinction" (p. 174). Contrary to this trend, however, Truss offered some very persuasive examples of the value of the hyphen. On a basic level, the hyphen can be "used to avoid an unpleasant linguistic condition called 'letter collision'" as evidenced by "de-ice" rather than "deice" or "shell-like" rather than "shelllike" (p. 173). Beyond this practical function, the hyphen can have a profound effect on meaning. Truss asked us to consider the difference in meaning between a "pickled-herring merchant" and a "pickled herring merchant" (p. 169) or between a "re-formed rock band" and a "reformed rock band" (p. 171). Perhaps the example with the greatest impact is the difference between "extra-marital sex" and "extra marital sex" (p. 168). Clearly the hyphen can be important!

Are there any guidelines for use of the hyphen? Going back to the Greek origins of the word, Truss (2003) informed us of the following:

> The phrase from which we derive the name hyphen means "under one" or "into one" or "together", so is possibly rather more sexy in its origins than we might otherwise have imagined from its utilitarian image today. Traditionally it joins together words, or words-with-prefixes, to aid understanding; it keeps certain other words neatly apart, with an identical intention. (p. 169)

What implications might this have for our discussion of the person-in-environment perspective in social work? The phrase is almost always hyphenated in the social work literature, but what is the effect of these hyphens? Do they aid our understanding by joining the words together or by keeping them apart? Do the hyphens encourage us to perceive connections between people and environments, or do they emphasize the distinction between two separate and mutually exclusive entities? Raising concerns about the hyphens is not simply idle wordplay; the issue has been raised as important in the social work literature.

Meyer (1976) wanted to erase the hyphens altogether because, in her opinion, they distorted the very nature of the interaction between people and environments: "the erasure of hyphens in person-in-situation (a linear concept) has been a conceptual goal in social work" (p. 201). Still pressing the case two decades later, Meyer (1995) observed that "over the years, its hyphenated construction has contributed to the continuing imbalance in emphasis on the person *or* the environment. The term itself suggests that the person and the environment are independent of each other" (p. 16). For Meyer, the hyphens served a negative function of separating the concepts of person and environment. Such separation resulted in social workers having to make a choice between working with the person or working with the environment, thereby "assigning peripheral status" (Meyer, 1995, p. 16) to the other. Meyer concluded that "by 1970, it was time to review and rethink the person-in-environment construct so that social workers would find it possible to intervene in a more *transactional* fashion in cases that were clearly (nonhyphenated) psychosocial events" (p. 18). Overall, the concern appeared to be that the hyphens in "person-in-environment" were contributing to a limiting linear notion of the concept. Person was central, with the environment relegated to the periphery. Erasing the hyphens might allow a broader understanding of the transactional relationship between people and environments.

Germain and Gitterman (1987) agreed and proposed this punctuation revision: "In the ecological perspective, the equation denoting person:environment relationships substitutes a colon for the hyphen to underscore their transactional nature and to signify repair of the former discrete person–situation relationship" (p. 489). Was this the resolution? Could the use of a colon broaden the concept and begin to express its transactional nature? Truss (2003) considered the colon to be "a kind of fulcrum" (p. 119). A fulcrum, like the central point of a teeter-totter in a playground, suggests shifting emphases and the possibility of balance. It appears that *person:environment* may be a more effective presentation of the transactional concept than *person-in-environment*. Yet somehow the term never caught on in the social work literature. Although it is a meaningful expression as a stand-alone term, perhaps it becomes too confusing or ambiguous when used in a sentence. Consider, as an example, this sentence: "We must accept the person:environment perspective is essential for our work." There are several possible meanings for this sentence depending upon one's assumptions about the role of the colon.

Allen-Meares and Lane (1987) also grappled with the limitations of the symbols commonly used to represent the relationship between people and environments. Rather than tackling the punctuation marks, they manipulated capitalization. They depicted early social work activities in settlement houses as Person-ENVIRONMENT to emphasize the priority attached to community development. Later emphasis on individual therapy and treatment was labelled as PERSON-Environment to reflect the concentration on individual growth and functioning. The current goal for social work was presented as PERSON-ENVIRONMENT to bring "back into focus the importance of the social and physical environments in the lives of people" (p. 515).

Although infrequent, there have been some efforts in the social work literature to challenge the conventional grammatical depiction of "person-in-environment." While these efforts have not yet succeeded in revising the long-standing phrase with new symbols, they are evidence at least of a discontent with the status quo and a warning that we may be perpetrating a limited notion of the relationship between people and their environments by the very symbols we use to represent that relationship. Maybe the problem is not just the arrangement of words and punctuation. Could there be more serious issues with the language itself?

LANGUAGE AND LIMITATIONS

Language has been described as a social construction (Arbib & Hesse, 2008; Gergen, 1985). According to Coale (1998), language is "a product of the interaction between environmental stimuli and what our nervous systems are expecting as mediated by culture.... Culture constrains us by linguistically shaping our beliefs about what is true, functional, right, workable, and moral" (p. 48). Foley (2005) emphasized that human language structures are learned through culture; they are neither innate nor universal. Once thought to be common to all languages, the distinction between nouns and verbs is not universal at all. Foley (2005) grouped the Western European languages as "noun–verb languages" (p. 58), but suggested another distinct group of exotic Aboriginal languages that are "precategorical" (p. 59) in that they are based on relationships rather than labels and hierarchies.

As I explained in the Introduction, I am limited to the English language. I have been told many times over the years by Aboriginal colleagues and friends that English is a fine language for labelling and ranking people, things, and events. The English language may not, however, be as useful for expressing relationships, spiritual connections, and holistic concepts. There have been many occasions where an Aboriginal teacher or colleague has tried to translate an Aboriginal concept into English for me, only to end up frustrated and saying something like: "I can only go so far in English. This is the best I can do. The whole idea isn't there, but there just aren't the English words for it." Many Aboriginal authors have risked sharing key concepts from their own languages and healing practices, roughly translated into English, so that others might begin the process of coming to know these holistic notions of healing and identity. I think it might be valuable to examine some of these concepts, as offered in the literature from Aboriginal healing, to better understand how the English language may be limiting our efforts to express connections with the natural environment.

There is a caution here. Castleden and Garvin (2004) offered a fascinating discussion of the difficulties with translating Indigenous concepts into English when they contrasted the Indigenous concept of "sacred land" with the Western academic concept of "therapeutic landscape." Both labels refer to locations with enduring reputations for healing, and the terms are often used interchangeably in Western academic discourse. Yet the authors claimed that "Sacred Land, as a viable and acceptable analytic concept, has been de-valued and its meaning fundamentally changed through its integration into Therapeutic Landscape" (p. 93). The Indigenous notion of sacred land implies horizontal or non-hierarchical relationships of equality, trust, and respect. This equality "dissolves in the transformation into the vertical concept of Therapeutic Landscape where human construction becomes part of what determines whether a land-scape is *therapeutic*" (p. 92).

Castleden and Garvin (2004) described this process as an academic colonization that diminishes the power of the original Indigenous concept in order to make it more acceptable and open to manipulation within Western academia. Four Arrows (Jacobs, 2006) presented such academic colonization as the "third wave of the attack" on Indigenous peoples (p. 20), following the first wave of violence and disease and the second wave of institutional control. Particular concern has been expressed about the subsequent reapplication of distorted concepts to Aboriginal communities in the design of service delivery that pretends to be culturally appropriate.

Heeding this warning, I do not intend to present terms from the Aboriginal social work and healing literature as if I control or fully understand their meanings and implications. As much as possible, I have used the original authors' own words to describe the terms in English, then I present my reflections on the limitations of the English language for full conveyance and understanding of the concepts.

From his Oglala Sioux perspective, McGaa (1990) explained how *Mother Earth* is at the core of Aboriginal healing processes, healing for the people and the natural environment. He considered the English term "ecology" to be much more limited than the notion of *Mother Earth*, who demands respect and protection in return for survival. Using a similar argument, Battiste and Henderson (2000) argued that the English term "nature" falls short of the Mi'kmaq expression *kisu'lk mlkikno'tim*, which translates more as "creation place" (p. 77). The natural world is understood as constantly transforming, a realm that must be respected and experienced through connections and relationships rather than detached study.

Colorado (1991), from her Oneida heritage, introduced us to the term *Gii Lai*, which translates into English as "the still quiet place" (p. 21). On one occasion, she elaborated this term to me as that calm pool just beyond the lip at the top of a waterfall. Here, in the *Gii Lai*, salmon who have exhausted themselves jumping up the waterfall can rest and renew themselves in the oxygen- and nutrient-rich deep-water pool before continuing their strenuous journey upstream. As she has explained in her writing, healing work involves guiding people to that still quiet place and working there with "the energies of the earth" (Colorado, 1991, p. 21).

I have had the privilege of being with Dr. Colorado as she has conducted healing work in northern Canada. I have watched her sit in silence with a person for an extended period of time before commencing the healing work (actually the healing work had already begun, I just couldn't recognize it). I would grow uncomfortable with the lengthy silence and apparent inactivity. In my Western system, the interview would commence at 3:00 p.m. if that was the scheduled appointment time. As Dr. Colorado later explained to me, she was "waiting for the spirits to show themselves." There would be no point in engaging in a healing dialogue until we were all in "the still quite place," the *Gii Lai*.

Timpson and Semple (1997) described a suicide-prevention initiative of the Shibogama First Nations Health Authority in northwestern Ontario. The program was named by a local elder as *Payahtakenemowin*, a term said to mean "peace of mind" in English (p. 98). In common English usage, "peace of mind" conveys a sense of internal contentment. I suspect *Payahtakenemowin* is a much broader notion since the described strategy involved community relationships, family care, and wilderness experiences.

Baskin (1997, 2005) similarly described the naming of a family-violence program using the Ojibway term *Mino-Yaa-Daa*. Commonly translated into English simply as "healing together," the term in its fullest sense includes broad and profound Aboriginal notions of healing through relationships that involve "the gifts of the earth" and "teaching respect for the self, family, community, and the earth" (Baskin, 2005, pp. 172–173). It seems the "together" of "healing together" includes much more than a social support group.

An Aboriginal justice diversion program in Winnipeg adopted the Ojibway name *Ganootamaage*, which translates into English as "speaking for" (Mallett, Bent, & Josephson, 2000, p. 61). Yet this is not "speaking for" in the sense of legal representation or advocacy. There is a strong spiritual component, as the offence itself is perceived as "broken-spirited relations" (pp. 62–63) between the offender and the victim. Through a community council, a holistic healing plan is developed that often involves restoring some form of balance with the physical environment as well as between the people. This can be attempted through the use of traditional hand-picked medicines from the land, or through lessons from Elders on "how an individual's mind, heart, body, and soul are connected to Mother Earth" (p. 72). *Ganootamaage* seems to go beyond "speaking for" to a notion of "standing with" in a spiritual sense, recognizing the healing powers of the natural environment.

Hart (2002) described the process of healing as a journey towards *Mino-pimatisiwin*, a Cree term that translates into English as "the good life" (p. 44). Once again, there are dangers here with the English translation, dangers arising from assumptions attached to our common English usage of the phrase. "The good life" is a label often used in English to describe the lifestyle of an individual who has achieved an existence of comfort with abundant material goods. Hart's concept of *Mino-pimatisiwin* is very different; it involves sharing, spirituality, connectedness, and respect for all things because we all came from the same Mother Earth (p. 46). Reconnection with the land "helps people to see life in a broader sense that incorporates both the physical and spiritual realms" (p. 49) and is essential for the journey towards *Mino-pimatisiwin*. Castellano (2006) translated the term *pimatisiwin* as "being alive well," a process that requires observance of all nature.

Bastien (2004) offered the Blackfoot term *Ao'tsisstapitakyo'p* when she described how knowledge is acquired through relationships with all things natural and spiritual, a coming-to-understanding that is profoundly connected with the natural world (p. 2). McGaa (1990) offered a similar notion with the Lakota Sioux expression *Mitakuye Oyasin*, which he translated into English as "we are related to all things" (p. xvii) or "we are all related" (p. 203). This vision of a relationship with all things goes far beyond Western notions of extended family to "the animal or plant world as our brothers and sisters, to Mother Earth and Father Sky and then above to *Wakan Tanka*, the *Unci/Tankashilah*, the Grandparent of us all" (p. 208).

In their description of a traditional framework for decision-making, Crowshoe and Manneschmidt (2002) used the Blackfoot term *Akak'stiman*, which they translated into English as "law-making" (p. viii). From the comprehensive process described in the book, however, I again sense a meaning that goes far beyond the conventional English-language notion of law-making whereby delegated authorities develop and approve standard rules for society. *Akak'stiman* appears to have a base in spiritual ceremony, in particular the annual Sun Dance. Here the Blackfoot people "witness the making and enforcement of their laws" (p. viii), which I understand to involve a process of receiving laws rather than just authorities imposing laws or directions for the people. Laws are discovered in the relationships of the people and the natural world, not created by experts.

Bastien (2004) also spoke of *Ihtsipaitapiiyo'pa*, which is "the great mystery that is in everything in the universe" and "lives in each and every form of creation" (p. 77). Battiste and Henderson (2000) suggested similar spiritual meanings for the Algonquian term *mntu*, the Iroquois word *orenda*, and the Lakota term *nigila*, which they translated roughly into

English as "that which dwells in everything" (p. 76), although they suggested that a fuller approximation would include a combination of the English words "dignity," "power," and "force" (p. 78). Cajete (2000) used the Navajo term *ho'zho'*, which translates into English as "natural beauty" but also involves "the Navajo notion of proper relationship between humans, nature, and each other" (p. 184). The expression goes beyond passive scenery and landscape to incorporate active interrelationships and responsibilities between human beings and the natural world.

Longclaws (2005) used the Anishnaabe term *oda aki*, which is translated into English as "centredness" (p. 362). There is a spiritual component of this centredness that goes beyond location in the middle. To Longclaws, *oda aki* involved "the achievement of balance—peace and harmony with oneself and all other living things" (p. 362) and was the goal of Aboriginal social work. There is a sense here of finding yourself in the universe, of actively strengthening relationships with the environment. Problems are understood "in terms of the spiritual relationship to the environment" (p. 366).

All of these Aboriginal terms from the healing literature call out for an appreciation of the mystery and energies that are present everywhere in an ever-changing natural environment. This direction appears consistent with the notions of stewardship and respect for the environment that we saw emerging in the literature from rural and northern social work practice, but there is an added dimension of spiritual connection with *Mother Earth* as we saw in the Aboriginal social work literature. Aboriginal languages tend to be verb-based (Cajete, 2000; Witherspoon, 1977), expressing the active and continuing act of co-creation with the Earth. We seem to have difficulty incorporating this spiritual dimension in any English-language approach to expressing the relationship between people and the environment.

Compared with the examples from Aboriginal terminology, it could be argued that English nouns tend to be relatively static labels for people, things, and events. How much of a problem might this language difference pose for our efforts to further our understanding of the relational aspects of person-in-environment? Four Arrows (Jacobs, 2006) considered that "even the most progressive environmental movements of Western culture are superficial" when compared with the "deep knowing embedded" in Aboriginal languages (p. 180). In the epilogue to his book *Unlearning the Language of Conquest*, Four Arrows gave this warning:

> The language of conquest ignores the Indigenous idea that we are all related and thus, to lie to one another is to lie to ourselves. Communication is a sacred release of power. Words can literally sing things into existence. A language of conquest has the power to bring about destruction. A language of truth has the power to renew. (Jacobs, 2006, p. 274)

At this point, I must pause to address a potential bias that may be apparent in my discussion of language and sense of place. When considering the limitations of the English language for expressing notions of holism, I have emphasized concepts from Aboriginal languages and knowledge systems as a potential source of new understandings for mainstream social work. This choice reflects my own experiences with Aboriginal colleagues and teachers, and my familiarity with the relatively new literature from Aboriginal social work. It is not my intention to advocate a full-scale adoption of an Aboriginal world view for mainstream social work. The process is not as simple as that. As Four Feathers has explained:

> No single race of people can lay claim to "Indigenous wisdom." It lives deep within the heart of every living creature. Anyone who remains deeply aware of the rhythms of the natural world can remember it. Unfortunately, it seems that most of us have lost or are losing this "primal awareness," largely because of the language of conquest. (Jacobs, 2006, p. 18)

In today's world, there is a wealth of vibrant cultures, world views, and languages available to challenge dominant ideologies and systems. Representing place-based experiences from many parts of the globe, these diverse traditions offer a multitude of concepts with potential for the work ahead of us in learning to live well in place. In portions of this book, I have shared some of my own learning and initial understandings from the Aboriginal teachers who have helped me. I trust that readers can draw on their own traditions, experiences, and learning to bring their own perspectives to the common task of creating new ecological understandings. There can be no single absolutely correct perspective on the environment. This is not a case of looking for the one culture that has it right; this is a case of seeking understanding from the wisdoms generated by many peoples' experiences in many places on the planet.

PREPOSITIONS AND PROPOSITIONS

This journey through language perspectives now brings us to a major obstacle for expressing the interactions between people and the world around us. As long as I accept "person-in-environment" as the foundation metaphor for social work practice, I am forced to separate the person from the natural environment and consider them as two distinct entities. Of course they are interacting, but they are still separate and clearly unequal. From a grammatical perspective, "person" is the subject and "in environment" is the modifier. Human activity is primary, with the environment as merely the setting or backdrop. This view is not consistent with the natural environment as I have experienced it (especially in northern Canada), or with the approaches of stewardship and respect for the natural environment that are present in the rural, remote, and Aboriginal social work literature and beginning to emerge in the mainstream literature.

Beyond suggestions for tinkering with punctuation and capitalization, there have been calls in the social work literature to reconsider the conventional notion of "person-in-environment." One of the clearest was Canda's (1998) demand for social work to revisit the person-in-environment concept "in a dramatic way" because the person is "not separable" from the natural environment (p. 103). But how have we in social work responded to such calls? As a profession rooted in the English language and with an academic preference for deconstruction and analysis, we have tended to respond to the challenge by trying to learn still more about either the "person" or the "environment." Like moths to a flame, we are attracted to the nouns. We continue to split these entities into further labelled component parts and subcategories for analysis.

The expression "person-in-environment" does contain two nouns, but they are joined by the preposition "in." What is the significance of this preposition? Often overlooked in our rush to focus on the nouns, the simple preposition may have more power than we think when it comes to expressing relationships between a person and his or her environment. Through our preferred process of deconstructing and refining our understanding of the nouns, we may be missing the sense of connection or relationship conveyed by the simple preposition "in."

My first hint of the relational power of prepositions came when I was reading a report prepared by the Canadian geographer Hamelin (1984) following a national seminar on public administration in the North. He distinguished between managers "being in the North" and managers "being of the North" (p. 180). Rather than concentrating on the nouns (presenting new categories of managers or more refined scales of nordicity), Hamelin totally rearranged the relationship between manager and northern community by simply changing the preposition. His distinction rang true for me from my own northern experience. I had vivid recall of managers I had encountered who were "in the North," temporarily living there but imposing southern models and methods to solve northern problems. I had also known managers who were "of the North," men and women who had been engaged with the region for a period of time and who respected its features, rhythms, vulnerabilities, and strengths. Hamelin had caught this meaningful relationship with place by changing a preposition.

Would the same substitution be helpful for overcoming the limitations of the person-in-environment concept? Could I simply replace "in" with "of" as Hamelin (1984) had done? What vision is conveyed by "person-*of*-environment"? Certainly, I think this is an improvement over "person-in-environment." There is a sense here that the person comes from the environment. There is also an aspect of possession, more in the sense of belonging than of ownership. People are not just *in* the natural environment. They may experience a sense of belonging or inclusion in certain places, a sense of being *of* that place. Yet there is still a problem. The person may be *of* the place, with expanded notions of identity and belonging, but they can still be separated from the place, and the implied relationship is in the past. The emphasis is on where the person came from, not where the person is functioning at the moment. What other alternatives might there be for challenging "in" as the relational preposition for the foundation metaphor upon which we have built our practice?

As we saw in Chapter 5, deep ecologists reject any division between people and the non-human world, and argue that human identity is rooted in an ecological consciousness, a "moving away from a view of person-in-environment to one of self as part of a 'relational total-field'" (Besthorn, 2001, p. 31). It seems to me that this notion of "relational total-field," as expressed in academic English, approximates the Aboriginal notions of *Ihtsipaitapiiyo'pa* (Bastien, 2004) or *mntu* (Battiste & Henderson, 2000) discussed earlier, the great mystery that dwells in everything. In addition to conveying the image through complicated nouns and adjectives, Besthorn (2001) also addressed the simple matter of the relational prepositions when he went on to explain that "rather than experiencing ourselves as separate from our environment and existing *in* it, we begin to cultivate the insight that we are *with* our environment" (p. 31).

Is this the answer? Clearly, living *with* our environment gives a profoundly different perspective than living *in* our environment or being *of* the environment. Interaction *with* another entity implies recognition of the identity and autonomy of the other. The natural environment ceases to be a lifeless backdrop, but emerges as a living partner and collaborator with whom we co-exist. This relationship can be read from either direction—"people *with* environment" or "environment *with* people." This perspective is helpful, but I am concerned that there is still a fundamental flaw. "People *with* environment" may be more inclusive, egalitarian, and respectful of the environment than "people *in* environment" or "people *of* environment," but it still assumes two separate and distinct entities. This expression continues to fall short of the spiritual connections between people and *Mother Earth* expressed earlier by the Aboriginal authors. According to Cajete (2000), "Indigenous people are people of place, and the nature of place is embedded in their language" (p. 74). Elsewhere, I have argued for understanding the person and the environment as one entity so that "the connections between them become more than social interaction or landscape modification; they are profound expressions of the same whole and assume a dimension of energy or spirit" (Zapf, 2005b, p. 70). This is a perspective of "person *as* environment."

The *Gaia* hypothesis put forward by Lovelock (1987) presented the Earth as a living entity, one living being organized to support life. Relating this Gaian perspective to social work, Penton (1993) promoted creative circular definitions of client problems, arguing that our conventional assessments assume a precision and linearity that do not exist in nature. Such circularity may run counter to prevailing notions of cause and effect in the research and practice communities. Lovelock himself has been criticized for using imprecise language or fuzzy concepts in advancing the Gaia hypothesis, yet Morito (2002) explained the following:

Lovelock is not engaged in poetry. He is attempting to find a language
in which to situate his analysis of the standard assumptions in science,
a language that would demarcate his theory from standard theories in
biology and the life sciences. (p. 74)

We continue to search for such language. The dominance of the
English language in the social work literature makes it difficult to avoid
disconnections between person and environment. Like author Margaret
Atwood (1995), we seek "the lost syllable for 'I' that did not mean separate" (p. 54).

DISFLUENCY AND ENVIRONMENTAL DISCOURSE

In a highly personal essay on coping with a speech impediment (stutter-
ing), Zalitack (2005) wrote of a pattern that she called "disfluency." She
described stressful situations where she would be afraid that she would
not be able to express clearly what she really wanted to say. On such anx-
ious occasions, she often fell back on old familiar patterns. She described
"[reverting] to my old habits of speaking too fast, and choosing easy words
instead of those I really want but fear I can't get out, a coping skill known
as disfluency" (p. 49). Zalitack explained that this often happened "in a
familiar setting with established routines and relationships" (p. 49). Where
history and role expectations called for ease of communication, she would
be reluctant to try new words or manners of speech, even if that ultimately
meant not being able to fully express what she really meant. Fear of failing, of stumbling, of looking incapable, all of this combined to cause her
to fall back on the easy words and patterns instead of taking the risk to
push forward with new ideas and strategies. I am indebted to Zalitack
for her honest discussion of these personal ordeals, and for introducing
me to this concept of disfluency.

It seems to me that disfluency has been a pattern in social work when
we have tried to speak or write in new ways about the physical environment but have soon reverted to old patterns and easy words. Now that I

understand the concept, I can recognize that I have experienced disfluency in my own professional work. As explained earlier in this chapter, I am aware that the words and expressions I have in English may not be adequate to capture some of the holistic notions of balance, belonging, and spirituality that arise when exploring relationships between people and the natural world. I have experienced the fear of sounding foolish or incompetent when I try to speak or write of these things. I have found myself more comfortable presenting on these issues at conferences on regional science or spirituality than I have been at mainstream conferences on social work practice and education. At social work conferences, I find it far less stressful to revert to conventional approaches and familiar topics. I can avoid the risk of looking foolish or, even worse, irrelevant. I can use my easy academic words but not say what I really mean. This is disfluency. It has been my experience that many conference themes tend to reinforce this approach by inviting papers on predetermined familiar topics, while at the same time establishing the language to be used in discussion. Why is it so difficult to discuss possible spiritual connections with the environment at mainstream social work conferences and in mainstream publications?

The literature offers a number of possible interrelated explanations for social work's historical reluctance to incorporate spiritual concepts and language in its foundation literature (Zapf, 2007). Many social workers are employed in government agencies. The historical separation of church and state in Western countries may then begin to suggest why issues of spirituality are so difficult to discuss in those settings. Religious and spiritual factors have often been considered in the Western helping professions as problems, impediments, or pathologies, rather than being seen as strengths or resources in a client's situation. With its concern for issues of power and oppression, the social work profession has understandably been very cautious about any appearance of consciously imposing a spiritual frame of reference on any group. For practitioners seeking to demonstrate professional competence with intervention techniques that are under their own control, the broad scope of spiritual practice and understandings can be a problem. Spiritual issues, possibly perceived as unscientific or difficult to categorize and use in practice, could be awkward for a profession that is seeking to improve its status as an evidence-based scientific discipline.

Ecological thinking is "wholistic, receptive, trustful, largely non-tampering, deeply grounded in aesthetic intuition" and guided by "the almost sensuous intuiting of natural harmonies on the largest scale" (Roszak, 1972, p. 400). As social workers, we may find ourselves uncomfortable trying to operate in what Holmes and Byrne-Armstrong (2007) have called "the middle space … the space between different knowledges" (p. 44). Morito (2002) said:

> To some extent, conceptual clarity is sacrificed when taking holistic/non-mechanistic elements into account in explanations of ecosystems processes … a certain fuzziness in how wholes are causally effective is unavoidable. We cannot model such causal relations on the behaviour of billiard balls nor predict, in a mathematically rigorous way, just how holistic causality operates. (p. 76)

There appear to be real historical and immediate pressures in mainstream social work that discourage the risk of honestly talking about spiritual connections between human beings and the natural world. Like a Trojan horse, our ecological metaphor of person-in-environment contains some serious limitations hidden within the very expression itself. The words and grammatical structure suggest a dominant–subordinate relationship between two separate and distinct entities. Conventional patterns of academic analysis appear to support this separation. All of these forces contribute to a pattern of disfluency whereby we revert to our easy words—our person-in-environment—in our profession's established settings for debate and discussion. We have seldom risked talking about our relationship with and responsibilities to the natural world in a meaning-ful way, even though environmental issues have risen to the top of the priority list in the larger society.

CHAPTER NINE
PEOPLE AS PLACE

> To tackle problems that are both social and ecological,
> we need place-based models.... Understanding the
> dynamic interaction between nature and society requires
> case studies situated in particular places and cultures.
> (FRIIBERGH WORKSHOP, 2000)

The first eight chapters of this book have explored how the physical environment has been characterized and expressed within the social work profession and a number of related disciplines. Moving beyond mainstream social work sources, we have discovered an emerging interest in the concept of place. This sense of place appears common to many efforts to understand and begin to address environmental issues. In this concluding chapter, I argue for incorporation of this notion of place into the foundation models and practice approaches of social work.

Beginning with an overview of the perspectives covered in previous chapters highlighting the importance of place, I make an argument for retiring person-in-environment as the foundation metaphor upon

179

which our practice is built. At the macro level, humankind faces global environmental threats to our continued existence. Individuals, families, and groups face challenges of living meaningful lives within the resource realities of diverse locations at the micro level. Social work would do well to consider a more dynamic metaphor of "people as place" for our work in the 21st century.

RECLAIMING THE ENVIRONMENT: RECOGNIZING PLACE AND STEWARDSHIP

In spite of the high priority assigned to environmental issues by scientists and the general public, we have seen how the physical environment has generally been neglected, ignored, or rendered irrelevant in the mainstream social work literature. With no rationale or substantial explanation offered, broad notions of the environment have consistently been narrowed to considerations of only the social environment. The natural world and the built environment, when visible at all, have been presented only as background, as a lifeless backdrop for human activities. Sometimes the physical environment has been included in the opening rhetoric of social work textbooks and then dropped from the subsequent practice models, diagrams, and assessment protocols.

We have seen a few genuine calls for inclusion of the physical environment in the mainstream social work literature. Reminders of the interdependence of physical and social environments have arisen from observations that environments hold subjective meanings for people. On occasion, social work has borrowed concepts from other disciplines to explore these meanings (territoriality, personal space, welcoming and hostile environments). Initial notions of proximate environments and located behaviours have led to preliminary discussions of the potential importance of a sense of place for social work. The mid-1990s forum discussed in Chapter 3 and the related ecological credo for social workers (Berger & Kelly, 1993) envisioned the integration of social justice and environmental justice concerns. Sustainability and protection of the environment were seen as necessary for human well-being and essential for new models of social work. Respect for the physical environment has led to more recent suggestions of an eco-social work that recognizes the interrelatedness of environmental, social, political, and economic issues. So far, however, this activity has not been very successful in altering the direction of the mainstream profession, possibly due to the tremendous momentum of the profession's declared world view of person-in-environment.

At the margins of the social work profession we have found more fully developed perspectives on the physical environment. Rural/remote social work emphasizes the importance of context and promotes practice that is context sensitive and community embedded. There is an appreciation here that geography affects both where and how people live. People experience and express an attachment to place, a sense of belonging. Places have meaning for people; identity is connected to place. Following on from this attachment to place in the rural/remote social work literature is a sense of responsibility for a healthy physical environment, a sense of stewardship.

This notion of stewardship is further developed in the new literature on spirituality and social work. There are signs of movement beyond the notion of spirituality as a characteristic or quality of the individual, movement towards a broader awareness of human beings in a relationship with nature. Humans seek meaning in the context of the natural environment. Deep ecology denies any separation between people and the natural environment. Such emerging ecological consciousness (the ecological self) carries with it a responsibility to care for the planet.

Aboriginal social work, built on traditional knowledge, offers a world view that integrates landscape, community, spirit, and self. Life is a process of finding and expressing one's place in the cosmos, in the natural world to which we all belong. Traditional knowledge brings together the sacred and daily life, with a strong emphasis on the land and concepts of place. Links between place and world view are to be found everywhere in people (our geopsyches) and in the environment (spiritual landscapes). Active stewardship and responsibility to the land, to our common "Mother Earth," are paramount.

From the perspective of international social work comes the notion of environmental citizenship. Working to replace current political and economic agendas of individual states with a global ecological agenda, environmental citizens are also concerned with issues of oppression and exclusion from environmental decision-making. Questions of rights are emerging at international levels. Is there a human right to a healthy and supportive environment? Does the natural world have the right to protection from degradation, pollution, and destruction? Global accounts link environmental development and sustainability with economic and social development (although we have seen this to be far less developed in the mainstream North American social work literature, where the physical environment generally continues to be ignored). International definitions and policy statements hint at multiple environments (including the natural world) and suggest that social work concern itself with active environmental responsibility through global coalitions to care for the planet, but our dominant framework and models of practice do not make it easy for social work to contribute in a meaningful way at this time.

Disciplines outside of social work offer intriguing perspectives and concepts from their experiences with connections between people and the physical environment. From the world of art we find visual expressions of encounters with physical landscapes and an historical record, preserved in visual images, of ways of thinking about the land. The cinematic option of telling a place rather than telling a story challenges our cultural notions of place as mere scenery. Music theorists explain the process of authoring space through narrativization, local performances, and expressions of local knowledge and rhythms. Music has also been connected with intentional movement through space through such processes as soundscapes, songlines, and musical pathways. Exploring the link between wine and place through the development of *terroir* wines, viticulture attempts to capture place value or "somewhereness" in a bottle.

From a more traditional academic perspective, the discipline of sociology has been working on an embodied sense of place, *habitus*, and its implications for local life opportunities. Psychology has been actively exploring interrelationships between environments and human behaviour, with an emphasis on multidisciplinary efforts to understand the meaning of places. Environmental design is not only involved with understanding place as a concept, but also with active placemaking to create liveable and sustainable communities.

Geography, especially human geography, has been concerned with the mutual influences between people and the planet. Geography has also been very involved with the coming together of space and meaning to create a sense of place. Here we find discussions of place attachment, belonging, identity, and insiders and outsiders. Sustainability is a major concern on the grand level of the planet and the individual level of learning to live sustainably in a place. Personal geographies explore the question "Where do I fit?" in the locality, the region, and the world.

Education has also been looking at the intersection of space and emotional significance to create places. This is partly a content issue and partly a practical concern for those attempting to design learning environments. Concepts of community allegiance and rootedness have led to approaches of place-based education, actively connecting students with local environmental and social issues. Teaching and modelling such approaches leads the profession to concepts of stewardship and global environmental citizenship. A concerned focus on people and place has given rise to a new vision of education for learning to live well in place, which is very different from the long-standing goal of achieving context-free credentials.

Preliminary discussions about learning to live well in place may not be easy. Acceptance of the notion of living well in place, apparently quite simple at first glance, actually requires a transformation of thought, action, and understanding. In previous chapters, we have seen some indications of the direction the transformation process may take. Some authors have spoken about dwelling rather than merely living somewhere, of embracing rather than simply occupying space, of inhabiting rather than merely residing on the planet. Others have signalled the need for new understandings by hyphenating familiar activities to come up with terms and phrases that suggest a process of unlearning and critical reflection, terms such as re-education, re-envisioning, recapturing, re-inhabiting, coming-to-understand, and learning-anew-to-be-at-home. From the Aboriginal social work teachings, we have been introduced to notions of the good life, the great mystery that dwells in everything, nature as creation place, gifts of the earth, and relationships with all things. Through such examples, we can see the English language itself straining to express holistic concepts through awkward constructions that cobble together new words from old.

Mathews (2005) expressed with considerable eloquence the shortcomings of the English language for dealing with notions of place when she described love of place as

> an aching love—a love for which our modern languages lack words. For this love of place is not like other loves, of people or animals, artifacts, activities, causes. A loved being or thing or idea is held by us, held in our arms, in our imagination; our love casts a glow around it. But a loved place holds *us*, even if it exists only in memory; it causes everything within it, including ourselves, to glow. A loved place is not encompassed by our love; we are encompassed, loved, breathed into life, by it. There is little recognition or articulation of this kind of relation between self and world in modern Western thought—little attention to categories which express the way the world *makes room* for us as opposed to the way we act on it, impose ourselves on it. But many

of us sense this accommodation, sense that we are indeed received, and feel a huge but nameless emotion in response.... Being nameless, we have no option but to treat it as of little consequence. (p. 5)

Those who would push us in new directions speak of placemaking, wayfinding, and earthkeeping. Some have combined familiar words in new ways to express place-based concepts such as ecological literacy, nature-centredness, attentive living in place, or local life opportunities. These holistic and relationship-based concepts are often fuzzy, intuitive, imprecise, and multidisciplinary—qualities that tend to be shunned by Western academia and professions. Rather than taking the risk of trying to express these new concepts, we frequently find it easier to retreat to our comfortable words and conventional frameworks, a defence that has been labelled "disfluency."

Within my own English-language, technology-oriented, Western cultural traditions, I can see that we have to some extent been cutting ourselves off from the rhythms of the natural world. Returning to Four Feathers' terminology (Jacobs, 2006), we may have trouble remembering these rhythms. Directed by the language of conquest, we may be losing our ecological awareness. This loss has been labelled "nature-deficit disorder" (Lysack, 2008), reflecting our increasing reliance on electronic channels rather than our own direct senses to approach and experience the natural world. Is it too late for us to relearn, to remember, and to re-inhabit our own places?

MODELS AND METAPHORS: RETIRING "PERSON-IN-ENVIRONMENT"

Can social work be a participant in the evolving multidisciplinary efforts towards building a sustainable future on this planet? How might we incorporate a creative and dynamic sense of place as a foundation of our profession? Can we find clear and understandable language to express new place-based models of practice? To begin answering these questions, I must start with a clarification of what is meant by a model of social work.

The clearest presentation I have found of a social work model was offered nearly 30 years ago by Germain and Gitterman (1980) in their book *The Life Model of Social Work Practice*. According to the authors, a practice model consists of four components: (1) a metaphor, (2) a social purpose, (3) conceptual frameworks, and (4) practice methods. Any social work model must be grounded in a foundational *metaphor*, a world view or paradigm, a statement of belief about how the world operates. From this metaphor follows the *social purpose* or desired state of affairs. The social purpose is our mission, what we are trying to accomplish through our efforts, the vision of a better world that arises from the metaphor. Once the goal is in sight then *conceptual frameworks* are necessary to make sense of daily events and activities consistent with the metaphor. Conceptual frameworks are sets of labels and ways of thinking that allow a worker to assess everyday situations to determine what changes are needed to achieve the desired state of affairs. Finally, we come to the *practice methods*: what we actually do, the strategies and actions we implement to accomplish the social purpose.

Some brief examples from history may help to illustrate how these components can interact as a model of practice (Garvin & Seabury, 1997). When the world was understood as a battleground for the forces of good and evil (commonly known as the "moral metaphor"), the social purpose was one of just consequences—rewarding good while controlling, isolating, punishing, or eradicating evil. Conceptual frameworks that allowed relevant distinctions to be made in everyday life included such concepts as saints and sinners, worthiness and unworthiness, witches and sorcerers. Practice methods included penitentiaries, cutting off limbs, burnings, and executions. A later metaphor presented the world as a place where healthy organisms fought against illness (commonly known as the "disease metaphor"). The social purpose here was to cure afflicted hosts of unwanted disease processes. On the physical side, conceptual frameworks involved germs, bacteria, and other disease agents, while on the social side they included poverty, addictions, violence, and similar social ills. The dominant practice method within the disease metaphor was a pattern of study–diagnosis–treatment (familiar to us as the "medical model").

It is apparent from the above examples that the underlying metaphor or world view is the foundation for all that follows. The mission, the categories for labelling and understanding the world, and the interventions are all consistent with the foundation metaphor. Within the profession of social work, the foundation metaphor for at least three decades has been person-in-environment. The world is seen as a place where organisms are constantly adapting to their surroundings. (This world view has often been labelled the "ecological metaphor" in the social work literature, although I have difficulty with that terminology because it obscures the common pattern of ignoring the physical environment to focus only on the social environment.) From the foundation metaphor of person-in-environment follows a social purpose involving balance between people and environmental demands, often described as "goodness-of-fit." Conceptual frameworks then label various manifestations of stress and instances of imbalance or poor fit in the interactions between people and their environments. Practice methods feature a circular and multi-level problem-solving approach involving stages of problem definition, assessment, intervention, and evaluation.

The person-in-environment metaphor has accomplished many positive things in social work. It has helped us to integrate various levels of practice (casework, group work, family work, organizational and community work) into one relatively unified profession. "Person-in-environment" emphasized the transactional nature of our work, beyond fixing individuals in isolation. We were encouraged to look at intergroup relations, at issues of oppression, racism, and empowerment. Yet there have also been disadvantages to the person-in-environment metaphor and it may be time to question its continued prominence.

The notion of goodness-of-fit or balance between people and potentially damaging systems has been challenged by many, including feminist social workers (Bricker-Jenkins, Hooyman, & Gottlieb, 1991) and structural social workers (Mullaly, 2006), as more supportive of the status quo than any real progressive change. As we have seen throughout this book, the profession's tendency to limit the broad notion of environment to focus only on social environments has seriously inhibited our ability to engage with pressing environmental concerns. It is now time to move on, to retire person-in-environment as the foundation metaphor and goodness-of-fit as the social purpose of social work for the 21st century. If we merely pause to add environmental concerns only at the level of our conceptual frameworks, we will again be thinking about ecology rather than thinking ecologically (Morito, 2002). Meaningful change to our model of social work must come at the basic level of the foundation metaphor and social purpose.

"PEOPLE AS PLACE": A FOUNDATION METAPHOR FOR SOCIAL WORK

How do we replace person-in-environment at the core of social work? Can we come up with a meaningful alternative to guide us into relevance for the environmental issues facing humans and the planet while not losing sight of our important goals related to individual and social development? Based on the recent multidisciplinary interest in a sense of place and place-based models, I propose "people as place" as a potential metaphor on which to develop models of practice responsive to the crucial environmental issues of the early 21st century.

The "person" of person-in-environment implies a focus on individuals. Yes, we have techniques for working with groups and communities, but individual practice has assumed overall priority. Groups and communities are collections of individuals, each attempting to adapt to his or her own unique set of circumstances. Present environmental threats, however, demand a communal response. Hawley (1986) warned us that a sustainable relationship with the natural world cannot be achieved through the actions of individuals working independently, "but by their acting in concert through an organization of their diverse capabilities, thereby constituting a communal system" (p. 3). He emphasized that the basic assumption of human ecology is that "adaptation is a collective rather than an individual process" (p. 12). Thinking ecologically involves thinking about "people" rather than the individual "person."

A case has already been made for the inadequacy of the term "environment" in a foundation metaphor for social work. The broad notion of environment has repeatedly and effectively been narrowed to the social environment in most of the mainstream social work literature. Environment has been perceived as separate from the human actors. Transactions are acknowledged between persons and environments, but these are exchanges between separate entities. Over time, this separation has led to a subordination of the environment relative to human concerns. The environment has been relegated to the background, a setting for human activities. If we replace "environment" with "place" in the metaphor, we overcome this troublesome separation. "Place" combines location and physical environment with character, meaning, and emotional significance for people; it is a multidisciplinary concept that brings together the natural world and human history, activities, and aspirations. "Place" is an interactive and holistic concept. Social components cannot easily be extracted from "place" for separate consideration (as was the case with "environment").

If the nouns "people" and "place" are accepted for the new metaphor then what is the most appropriate connector? Use of "in" in the person-in-environment metaphor has reinforced the separation of person from environment, situating the person as primary with the environment as secondary or supporting. In Chapter 8, we explored several alternatives. The preposition "of" was considered because it introduces a sense of belonging. A better option was "with," which suggests equality in the relationship but unfortunately still assumes two separate entities. After looking to the writings of Aboriginal authors, I concluded that "as" may be the most appropriate word in the English language for our purposes. An expression of "people *as* place" conveys a unity and holism that brings us immediately to concerns of sustainability and stewardship.

Viewed within a paradigm or metaphor of people as place, humans cannot be understood as separate from the natural world. A major objective for social work at this time, according to Coates (2003), is to "help bring about a transformation of society into one with a vision and mandate that recognizes that we are intimately and symbiotically connected with nature and all people" (p. 97). Human health and welfare are bound up with environmental health and welfare. Environments are not merely lifeless backdrops for human activity, any more than people are merely temporary actors in an ongoing natural system. We are entwined with the natural world in a continuing process of co-creation. Human development cannot be separated from stewardship of the earth. In short, we are our surroundings. People as place.

SOCIAL PURPOSE: LIVING WELL IN PLACE

If people as place is accepted as the foundation metaphor, what might resulting social work models look like? We know that a social purpose, a desired state of affairs, follows from the metaphor. If the world is understood as a process of continual co-creation involving people and the natural world then what might be the ideal vision of a better place? What are we trying to accomplish? The social purpose of such an ecological model could be "living well in place." As we saw in Chapter 8, some preliminary work has already been done on this notion of "living well in place" in other disciplines, primarily education (Orr, 1992). Many aspects have already been identified: living well ecologically, living well politically, living well economically, living well spiritually, and living well in community (Haas & Nachtigal, 1998). This idea of living well in place integrates social justice with environmental justice, human rights with environmental rights, and human responsibilities with environmental responsibilities.

Aspirations to living well in place can be applied at many levels of existence. How do I live well within myself (my health, my internal environment)? How do I live well in my home, in my neighbourhood, in my region, in my country, on my planet, in the universe? Obviously, dynamic sustainability is a key factor of living well in place; otherwise, we risk losing the very space to which our meanings, identity, and survival are attached. Living well in place is a process and not an end state.

IMPLICATIONS

With a foundation metaphor of "people as place" and a social purpose of "living well in place," new ecological models of social work practice will require conceptual frameworks for making sense of everyday activities and situations. This is the next task and it will be exciting work. Some concepts from other disciplines, cultures, and languages have been introduced in this book. Many others remain to be discovered and explored. That work, however, is beyond the scope of this book. My intent here was to identify the shortcomings of our current dominant social work model and make a case for adopting person as place as the foundational metaphor for the next generation.

Readers seeking a complete how-to manual for ecological social work may be disappointed, but we are simply not there yet. Shifting the momentum of mainstream social work is a huge task that must begin with informed commitment to a new world view and sense of purpose. If we accept person as place as the foundation metaphor and living well in place as our social purpose then we can move on to what Berkes (2002) called "local observations and place-based research" (p. 336). From these activities, we will develop the concepts and language useful for our work.

Those social workers looking for something more immediate could do well to start with the ecological credo for social workers (Berger & Kelly, 1993) and the suggested environmental actions (Berger, 1995) that were presented in Chapter 3.

Within the new metaphor, there will be room for many practice methods. As social workers, we will continue working with people to accomplish change at various levels. As our ability to think ecologically develops, we may find some methods giving way to newer and more appropriate practices. Effective strategies for accomplishing environmental change may emerge, but essentially we will always be working with people and using our skills for communication and influence. The difference will be in the underlying world view, the understanding of the unity of people and the natural environment. From a metaphor of people as place, life may be perceived as more of a journey or quest, a process of placemaking and wayfinding, rather than a series of problems to be solved.

Social critic and comedian George Carlin (2004) said this about place:

> I've noticed that when people speak these days, location seems important to them; and one location in particular: *there*. They say such things as *don't go there; been there, done that;* and *you were never there for me*. They don't say much about *here*. If they do mention here, they usually say, *"I'm outta here."* Which is really an indirect way of mentioning there, because, if they're outta here, then they must be going there, even though they were specifically warned not to. It seems to me that here and there present an important problem because, when you get right down to it, those are the only two places we have. Which, of course, is really neither here nor there. (p. 73)

Here and there—the only two places we have. This observation reminds me of the insider/outsider distinction we examined from phenomenological geography and the person/environment distinction we have endured in social work. Angus (1991) offered a fascinating Canadian perspective on here and there. He argued that the Canadian struggle for identity in relation to nature reveals a fundamental tension between home-making and wilderness (p. 20). Historically, Canadians have focused on home-making, on building structures for protection from a wilderness that must be subdued or controlled. We have focused on the *inside*, on the cultural and the social, on the tame and the safe, on the *here*. (I think this is exactly what we have done in social work as well. We have focused on the social environment where we are comfortable operating, our professional home, our *here*.) Angus (1991) warned, however, that such a focus on home leads us to a goal of only saving ourselves, a goal that is no longer sustainable. If we are to survive we must leave the comfort of here and rediscover our relationship with *there* (the outside, the wilderness, nature). How do we do this? How do we begin to instill a sense of ecological sensibility in our profession and our work?

With reference to young children, Pearce (1991) proclaimed the importance of children's literature as a foundation for "re-imagining ourselves back to nature" (p. 123). For children's books to assume this important transformation role, however, he said the following:

> [T]hey must convey more than information about environmental problems. They need to be rooted in an ecological sensibility that conceives of people as part of and interdependent with the rest of nature (and in more than a strictly biological way). (p. 123)

Our transformative task in social work is similar to that which Pearce has described. Our professional values and models of practice are conveyed through our literature base. We may be aware of and want to be part of the communal response to pressing environmental issues, but our literature base does not yet reflect the requisite ecological sensibilities. We will not succeed in "re-imagining ourselves back to nature" until our models and literature base have been transformed by ecological thinking.

The first step in such a transformation is a deep questioning or unlearning of our present assumptions.

My generation of social work practitioners and builders of knowledge has operated for 30 years with models of practice based on a metaphor of person-in-environment. Yet there have been great changes over that period. We are now confronted with the formidable "inconvenient truth" (Gore, 2006) of climate change and other environmental threats to our very existence, threats that were not considered or even anticipated when we built our practice models. Those models are no longer adequate for coping with the challenges faced by today's societies and the planet itself. Coates (2003) put things very clearly: "Social work has the choice of continuing to support a self-defeating social order or recreating itself to work toward a just and sustainable society" (p. 159).

Throughout this book, I have attempted to acknowledge the deficits of my generation's person-in-environment approach to social work. Building on environmental insights and activities at the margins of our profession and other disciplines, I have suggested a direction for the next stages of model building based on a foundation of people as place. It falls to the next generation of social work practitioners and theorists to complete this work. As social workers, they will be a transitional generation learning from many cultures and disciplines while they work to develop place-based models that can contribute to larger communal efforts for sustainability. As environmental citizens, they will be learning and teaching how to live well in place. Much depends on their work.

REFERENCES

Adams, M. (2006). *Sex in the snow: The surprising revolution in Canadian social values* (10th anniversary edition). Toronto: Penguin Canada.

Agnew, J.A., & Duncan J.S. (1989). Introduction. In J.A. Agnew, & J.S. Duncan (Eds.), *The power of place: Bringing together geographical and sociological imaginations* (pp. 1–8). Boston: Unwin Hyman.

Ahmed, Z. (2006). Stay out, India tells toxic ship. *BBC News* (January 6). Retrieved October 21, 2008, from news.bbc.co.uk/2/hi/south_asia/4588922.stm.

Alle-Corliss, L., & Alle-Corliss, R. (1999). *Advanced practice in human service agencies: Issues, trends and treatment perspectives*. Belmont: Wadsworth.

Allen-Meares, P., & Lane, B.A. (1987). Grounding social work practice in theory: Ecosystems. *Social Casework, 68*(9), 515–521.

American Psychiatric Association. (1994). *Diagnostic and statistical manual of mental disorders* (4th ed.). Washington: American Psychiatric Association.

Angus, I. (1991). Canadian roots for an ecological relation of self and world. In R. Lorimer, M. M'Gonigle, J.P. Reveret, & S. Ross (Eds.), *To see ourselves/to save ourselves: Ecology and culture in Canada* (pp. 1–22). Montreal: Association for Canadian Studies..

Arbib, M.A., & Hesse, M.B. (2008). *The construction of reality*. Cambridge: Cambridge University Press.

Architecture for Humanity (Eds.) (2006). *Design like you give a damn: Architectural responses to humanitarian crises*. London: Metropolis Books (Thames & Hudson).

Ashworth, G.J., & Graham, B. (Eds.) (2005). *Senses of place: Senses of time.* Aldershot: Ashgate.

Atwood, M. (1995). Marsh languages. In M. Atwood, *Morning in the burned house* (pp. 54–55). Toronto: McClelland & Stewart.

Barton, H. (Ed.) (2000). *Sustainable communities: The potential for eco-neighbourhoods*. London: Earthscan.

Bartuska, T.J. (1994). The fitness test: Building as a response to human-environmental factors. In T.J. Bartuska, & G.L. Young (Eds.), *The built environment: A creative enquiry into design and planning* (pp. 157–171). Menlo Park: Crisp Publications.

Baskin, C. (1997). Mino-Yaa-Daa: An urban community based approach. *Native Social Work Journal, 1*(1), 55–67.

Baskin, C. (2005). Mino-Yaa-Daa: Healing together. In K. Brownlee, & J.R. Graham (Eds.), *Violence in the family: Social work readings and research from northern and rural Canada* (pp. 170–181). Toronto: Canadian Scholars' Press Inc.

Basso, K.H. (1996). *Wisdom sits in places: Landscape and language among the Western Apache*. Albuquerque: University of New Mexico Press.

Bastien, B. (2004). *Blackfoot ways of knowing: The worldview of the Siksikaitsitapi*. Calgary: University of Calgary Press.

Battiste, M. & Henderson, J.Y. (2000). *Protecting Indigenous knowledge and heritage: A global challenge*. Saskatoon: Purich.

Bauder, H. (2001). Culture in the labor market: Segmentation theory and perspectives of place. *Progress in Human Geography, 25* (1), 37–52.

Bechtel, R.B (Ed.) (2008). Aims and scope. *Environment and Behavior.* Retrieved August 13, 2008, from www.sagepub.com/journalsProdAims.nav?prodId=Journal200783.

Bell, M.M. (1998). *An invitation to environmental sociology.* Thousand Oaks: Pine Forge Press.

Bender, L., David, L., & Burns, S.Z. (Producers), & Guggenheim, D. (Director). (2006). *An inconvenient truth* [Motion Picture Documentary]. United States: Paramount Classics and Participant Productions.

Berger, R.M. (1995). Habitat destruction syndrome. *Social Work, 40*(4), 441–443.

Berger, R.M., & Kelly, J.J. (1993). Social work in the ecological crisis. *Social Work, 38*(5), 521–526.

Berkes, F. (2002). Epilogue: Making sense of Arctic environmental change? In I. Krupnik, & D. Jolly (Eds.), *The earth is faster now: Indigenous observations of Arctic environmental change.* (pp. 334–349). Fairbanks: Arctic Research Consortium of the United States.

Besthorn, F.H. (2001). Transpersonal psychology and deep ecological philosophy: Exploring linkages and applications for social work. In E.R. Canda, & E.D. Smith (Eds.), *Transpersonal perspectives on spirituality in social work* (pp. 23–44). New York: Haworth Press.

Besthorn. F.H. (2002). Radical environmentalism and the ecological self: Rethinking the concept of self-identity for social work practice. *Journal of Progressive Human Services, 13*(1), 53–72.

Besthorn, F.H., & McMillen, D.P. (2002). The oppression of women and nature: Ecofeminism as a framework for an expanded ecological social work. *Families in Society: The Journal of Contemporary Human Services, 83*(3), 221–232.

Bishop, A. (2002). *Becoming an ally: Breaking the cycle of oppression in people* (2nd ed.). Halifax: Fernwood.

Bodor, R., Green, R., Lonne, B., & Zapf, M.K. (2004). 40 degrees above or 40 degrees below zero: Rural social work and context in Australia and Canada. *Rural Social Work, 9*, 49–59.

Bohm, P.E. (2005). Environmental issues. In F.J. Turner (Ed.), *Encyclopedia of Canadian social work* (pp. 122–123). Waterloo: Wilfrid Laurier University Press.

Bohmrich, R.C. (2006, January 15). The next chapter in the terroir debate. *Wine Business Monthly*. Retrieved May 16, 2007, from www.winebusiness.com/html/MonthlyArticle.cfm?dataID=42095.

Bone, R.M. (2002). *The regional geography of Canada* (2nd ed.). Don Mills: Oxford University Press.

Bonnes, M., & Bonaiuto, M. (2002). Environmental psychology: From spatial–physical environment to sustainable development. In R.B. Bechtel, & A. Churchman (Eds.), *Handbook of environmental psychology* (pp. 28–54). New York: John Wiley & Sons.

Booth, A. (1997). An overview of eco-feminism. In A. Wellington, A. Greenbaum, & W. Cragg (Eds.), *Canadian issues in applied environmental ethics* (pp. 330–351). Peterborough: Broadview Press.

Booth, A., & Jacobs, H.M. (1993). Ties that bind: Native American beliefs as a foundation for environmental consciousness. *Environmental Ethics, 12*, 27–43.

Bourdieu, P. (1990). Droit et passé-droit: Le champ des pouvoirs territoriaux et las mise en oeuvre des reglements. *Actes de las Recherche en Sciences Sociales, 81/81*, 86–96. (English translation of definition of "habitus" appears in Hillier & Rooksby, 2005, p. 21).

Brassard, A. (2002). *An integrated approach to creating community through safety planning processes*. Unpublished master's thesis, University of Calgary, Faculty of Environmental Design.

Bricker-Jenkins, M., Hooyman, N.R., & Gottlieb, N. (Eds.) (1991). *Feminist social work practice in clinical settings*. Newbury Park: Sage.

Brown, J.C. (Ed.) (1933). *The rural community in social casework*. New York: Family Welfare Association of America.

Brownlee, K., Delaney, R., & Graham, J.R. (Eds.) (1997). *Strategies for northern social work practice*. Thunder Bay: Lakehead University Centre for Northern Studies.

Brownlee, K., Sellick, M., & Delaney, R. (Eds.) (2001). *Social work with rural and northern organizations*. Thunder Bay: Lakehead University Centre for Northern Studies.

Bruyere, G. (1999). The decolonization wheel: An aboriginal perspective on social work practice with aboriginal peoples. In R. Delaney, K. Brownlee, & M. Sellick (Eds.), *Social work in rural and northern communities* (pp. 170–181). Thunder Bay: Lakehead University Centre for Northern Studies.

Bullard, R. (Ed.) (1993). *Confronting environmental racism: Voices from the grassroots*. Boston: South End Press.

Bullard, R. (1994). *Unequal protection: Environmental justice and communities of color*. San Francisco: Sierra Club.

Bullis, R.K. (1996). *Spirituality in social work practice*. Washington: Taylor & Francis.

Cajete, G. (1994). *Look to the mountain: An ecology of indigenous education*. Durango: Kivaki Press.

Cajete, G. (1999). Look to the mountain: Reflections on Indigenous ecology. In G. Cajete (Ed.), *A people's ecology: Explorations in sustainable living* (pp. 1–20). Santa Fe: Clear Light Publishers.

Cajete, G. (2000). *Native science: Natural laws of interdependence*. Santa Fe: Clear Light Publishers.

Calliou, S. (1995). Peacekeeping actions at home: A medicine wheel model for a peacekeeping pedagogy. In M. Battiste, & J. Barman (Eds.), *First Nations education in Canada: The circle unfolds* (pp. 47–72). Vancouver: University of British Columbia Press.

Canadian Association of Social Workers (CASW). (2000, March). *CASW national scope of practice statement (Approved by CASW board, March 2000)*. Retrieved November 21, 2008, from www.casw-acts.ca.

Canadian International Development Agency (CIDA). (1987). *Sharing our future: Canadian international development assistance*. Hull: CIDA.

Canda, E.R. (1988). Spirituality, diversity, and social work practice. *Social Casework, 69*(4), 238–247.

Canda, E.R. (1998). Afterword: Linking spirituality and social work: Five themes for innovation. In E.R. Canda (Ed.), *Spirituality in social work: New directions* (pp. 97–106). New York: Haworth Press.

Canda, E.R. (2008). Spiritual connection in social work: Boundary violations and transcendence. *Journal of Religion & Spirituality in Social Work, 27*(1–2), 25–40.

Canda, E.R., & Furman, L.D. (1999). *Spiritual diversity in social work practice: The heart of helping.* New York: Free Press.

Canda, E.R., & Smith, E.D. (Eds.) (2001). *Transpersonal perspectives on spirituality in social work.* New York: Haworth Press.

Canning, P., & Strong, C. (2002). Children and families adjusting to the cod moratorium. In R.E. Omer (Ed.), *The resilient outport: Ecology, economy, and society in rural Newfoundland.* (pp. 319–341). St. John's: ISER Books (Institute of Social and Economic Research, Memorial University of Newfoundland).

Cannon. T. (1994). Vulnerability analysis and the explanation of 'natural' disasters. In A. Varley (Ed.), *Disasters, development, and environment* (pp. 13–30). West Sussex: John Wiley & Sons.

Caragata, L., & Sanchez, M. (2002). Globalization and global need: New imperatives for expanding international social work education in North America. *International Social Work, 45*(2), 217–238.

Carlin, G. (2004). *When will Jesus bring the pork chops?* New York: Hyperion.

Carniol, B. (2005). *Case critical: Social services and social justice in Canada* (5th ed.). Toronto: Between the Lines.

Carroll, M.M. (1997). Spirituality and clinical social work: Implications of past and current perspectives. *Arete, 22*(1), 25–34.

Carroll, M.M. (2001). Conceptual models of spirituality. In E.R. Canda, & E.D. Smith (Eds.), *Transpersonal perspectives on spirituality in social work* (pp. 5–21). New York: Haworth Press.

Cascio, T. (1998). Incorporating spirituality into social work practice: A review of what to do. *Families in Society: The Journal of Contemporary Human Services, 79*(5), 523–531.

Cascio, T. (1999). Religion and spirituality: Diversity issues for the future. *Journal of Multicultural Social Work, 7*(3/4), 129–145.

Castellano, M. (2006). *Ethics and excellence in Aboriginal research.* The Honourable Justice Michael O'Byrne/AHFMR Lecture in Law, Medicine, and Ethics (April 16). University of Calgary.

Castleden, H., & Garvin, T. (2004). Therapeutic landscape as sacred land: Wolf in sheep's clothing? In J. Oakes, R. Riewe, Y. Belanger, S. Blady, K. Legge, & P. Wiebe (Eds.), *Aboriginal cultural landscapes* (pp. 86–96). Winnipeg: Aboriginal Issues Press (University of Manitoba).

Chan, C.L.W., & Ng, S.M. (2004). The social work practitioner–researcher–educator: Encouraging innovations and empowerment in the 21st century. *International Social Work, 47*(3), 312–320.

Chappell, R. (2006). *Social welfare in Canadian society* (3rd ed.). Toronto: Nelson.

Chaskin, R.J. (1997). Perspectives on neighborhood and community: A review of the literature. *Social Service Review, 71*(4), 521–547.

Chatwin, B. (1987). *The songlines.* New York: Penguin Books.

Cheers, B. (1985). Aspects of interaction in remote communities. *Australian Social Work, 38*(3), 3–10.

Cheers, B. (1998). *Welfare bushed: Social care in rural Australia.* Aldershot: Ashgate.

Cheers, B. (2004). The place of care—rural human services on the fringe. *Rural Social Work, 9,* 9–22.

Cheers, B., & Taylor, J. (2005). Social work in rural and remote Australia. In M. Alston, & J. McKinnon (Eds.), *Social work: Fields of practice* (2nd ed.) (pp. 237–248). Melbourne: Oxford University Press.

Clark, J.K., & Stein, T.V. (2003). Incorporating the natural landscape within an assessment of community attachment [Special issue of selected papers from the 2000 International Symposium on Society and Resource Management held in Bellingham, WA]. *Forest Service, 49*(6), 867–876.

Coale, A.W. (1998). *The vulnerable therapist: Practicing psychotherapy in an age of anxiety.* New York: Haworth Press.

Coates, J. (2003). *Ecology and social work: Toward a new paradigm*. Halifax: Fernwood.

Coates, J., Graham, J., Swartzentruber, B., & Ouellette, B. (Eds.) (2007). *Canadian social work and spirituality: Current readings and approaches*. Toronto: Canadian Scholars' Press Inc.

Coates, K.S. (2004). *A global history of Indigenous peoples: Struggle and survival*. New York: Palgrave Macmillan.

Cock, J. (1991). Going green at the grassroots: The environment as a political issue. In J. Cock, & E. Kock (Eds.), *Going green: People, politics, and environment in South Africa*. Cape Town: Oxford University Press.

Cohen, L. (1993). *Stranger music: Selected poems and songs*. Toronto: McClelland & Stewart.

Collier, K. (1984, 1993, 2006). *Social work with rural peoples* (1st, 2nd, 3rd eds.). Vancouver: New Star.

Colorado, P. (1991). A meeting between brothers—indigenous science. Interview with J. Carroll. *Beshara, 13*, 20–27.

Compton, B.R., Galaway, B., & Cournoyer, B.R. (2005). *Social work processes* (7th ed.). Belmont: Brooks/Cole.

Corey, M.S., & Corey, G. (2007). *Becoming a helper* (5th ed.). Belmont: Brooks/Cole.

Cornely, S.A., & Bruno, D.D. (1997). Brazil. In N. Mayadas, T. Watts, & D. Elliott (Eds.), *International handbook on social work theory and practice*. Westport: Greenwood Press.

Cornett, C. (1992). Toward a more comprehensive personology: Integrating a spiritual perspective into social work practice. *Social Work, 37*(2), 101–102.

Cossom, J. (2002). Working with informal helpers. In F.J. Turner (Ed.), *Social work practice: A Canadian perspective* (2nd ed.) (pp. 389–405). Toronto: Pearson Education Canada.

Crowshoe, R., & Manneschmidt, S. (2002). *Akak'stiman: A Blackfoot framework for decision-making and mediation processes*. University of Calgary Press.

Csiernik, R., & Adams, D.W. (2003). Social work students and spirituality: An initial exploration. *Canadian Social Work, 5*(1), 65–79.

Cummins, B., & Whiteduck, K. (1998). Towards a model for identification and recognition of sacred sites. In J. Oakes, R. Riewe, K. Kinew, & E. Maloney (Eds.), *Sacred lands: Aboriginal world views, claims, and conflicts* (pp. 3–14). Edmonton: Canadian Circumpolar Institute (University of Alberta).

Daley, M.R., & Avant, F.L. (2004). Rural social work: Reconceptualizing the framework for practice. In T.L. Scales, & C.S. Streeter (Eds.), *Rural social work: Building and sustaining community assets* (pp. 34–42). Belmont: Brooks/Cole.

Dawe, K. (2004). 'Power-geometry' in motion: Space, place and gender in the lyra music of Crete. In S. Whiteley, A. Bennett, & S. Hawkins (Eds.), *Music, space and place: Popular music and cultural identity* (pp. 55–65). Aldershot: Ashgate.

de Blij, H.J., & Muller, P.O. (1994). *Geography: Realms, regions, and concepts* (7th ed.) New York: John Wiley & Sons.

de Blij, H.J., & Murphy, A.B. (2003). *Human geography: Culture, society, and space* (7th ed.). New York: John Wiley & Sons.

Delaney, R. (2005). The philosophical and value base of Canadian social welfare. In J.C. Turner, & F.J. Turner (Eds.), *Canadian social welfare* (5th ed.) (pp. 13–27). Toronto: Pearson Education Canada.

Delaney, R., & Brownlee, K. (Eds.) (1995). *Northern social work practice.* Thunder Bay: Lakehead University Centre for Northern Studies.

Delaney, R., & Brownlee, K. (Eds.) (2009). *Northern and rural social work practice: A Canadian perspective.* Thunder Bay: Lakehead University Centre for Northern Studies.

Delaney, R., Brownlee, K., & Sellick, M. (Eds.) (1999). *Social work with rural and northern communities.* Thunder Bay: Lakehead University Centre for Northern Studies.

Delaney, R., Brownlee, K., & Zapf, M.K. (Eds.) (1996). *Issues in northern social work practice.* Thunder Bay: Lakehead University Centre for Northern Studies.

Deloria, V., Jr. (1999). *For this land: Writings on religion in North America.* New York: Routledge.

Devall, B. (1980). The deep ecology movement. *Natural Resources Journal*, 20(1), 299–322.

Devall, B. (1995). The ecological self. In A. Drengson, & Y. Inoue (Eds.), *The deep ecology movement: An introductory anthology* (pp. 101–123). Berkeley: North Atlantic.

DeYoung, R. (1999). Environmental psychology. In D.E. Alexander, & R.W. Fairbridge (Eds.), *Encyclopedia of Environmental Science* (p. 233). Hingham: Kluwer Academic Publishers.

Dezerotes, D. (2006). *Spiritually oriented social work practice*. Boston: Allyn & Bacon.

Diers, J. (2004). *Neighbor power: Building community the Seattle way*. Seattle: University of Washington Press.

Draper, D., & Reed, M.G. (2005). *Our environment: A Canadian perspective* (3rd ed.). Toronto: Nelson.

Drengson, A., & Inoue, Y. (Eds.) (1995). *The deep ecology movement: An introductory anthology*. Berkeley: North Atlantic.

Drouin, H.A. (2002). Spirituality in social work practice. In F.J. Turner (Ed.), *Social work practice: A Canadian perspective* (2nd ed.) (pp. 33–45). Toronto: Prentice Hall.

Drover, G. (1984). Policy and legislative perspectives. *The Social Worker, 52*(1), 6–10.

Drover, G. (2000). Redefining social citizenship in a global era [Special issue—social work and globalization]. *Canadian Social Work Review, 2*(1), 29–49.

Drover, G. (2005). Globalization and social welfare. In J.C. Turner, & F.J. Turner (Eds.), *Canadian social welfare* (5th ed.) (pp. 472–484). Toronto: Pearson Education Canada.

Drover, G., & MacDougall, G. (2002). Globalization and social work practice. In F.J. Turner (Ed.), *Social work practice: A Canadian perspective* (2nd ed.) (pp. 96–115). Toronto: Pearson Education Canada.

Drucker, D. (2003). Whither international social work? A reflection. *International Social Work, 46*(1), 53–81.

Eliot, T.S. (1971). Little gidding. In T.S. Eliot, *The complete poems and plays, 1909–1950*. New York: Harcourt, Brace, & World.

Ellerby, L. (2000). Striving towards balance: A blended treatment/healing approach with Aboriginal sexual offenders. In J. Proulx, & S. Perrault (Eds.), *No place for violence: Canadian Aboriginal alternatives* (pp. 78–98). Halifax: Fernwood Publishing and RESOLVE (Research and Education for Solutions to Violence and Abuse).

Entrikin, J.N. (1997). The betweenness of place. In T. Barnes, & D. Gregory (Eds.), *Reading human geography: The poetics and politics of inquiry* (pp. 299–314). London: Arnold.

Ersing, R.L., & Otis, M.D. (2004). An asset-based approach to promoting community well-being in a rural county. In T.L. Scales, & C.L. Streeter (Eds.), *Rural social work: Building and sustaining community assets* (pp. 160–177). Belmont: Brooks/Cole.

Evans, G.W. (1996). Environmental psychology as a field within psychology. *IAAP newsletter (International Association of Applied Psychology)*, 8(2), 4.

Evergreen Foundation. (1994). *Welcoming back the wilderness: A guide to school ground naturalization.* Toronto: Prentice-Hall.

Evernden, N. (1985). *The natural alien: Humankind and environment.* Toronto: University of Toronto Press.

Faiver, C., Ingersoll, R.E., O'Brien, E., & McNally, C. (2001). *Explorations in counseling and spirituality: Philosophical, practical, and personal reflections.* Belmont: Wadsworth.

Falk, R. (1994). The making of global citizenship. In B. van Steenbergen (Ed.), *The condition of citizenship.* London: Sage.

Farley, O.W., Griffiths, K.A., Skidmore, R.A., & Thackeray, M.G. (1982). *Rural social work practice.* New York: Free Press.

Feehan, K., & Hannis, D. (Eds.) (1993). *From strength to strength: Social work education and Aboriginal people.* Edmonton: Grant MacEwan Community College.

Fekete, J. (2007, March 8). Tories unveil vision of green Alberta. *Calgary Herald*, p. A1, A7.

Finnegan, R. (1989). *Hidden musicians: Music-making in an English town.* Cambridge: Cambridge University Press.

Fitzgerald, M. (1994). Environmental education in Ethiopia: A strategy to reduce vulnerability to famine. In A. Varley (Ed.), *Disasters, development, and environment* (pp. 125–138). West Sussex: John Wiley & Sons.

Foley, M.S. (2001). Spirituality as empowerment in social work practice. In R. Perez-Koenig, & B. Rock (Eds.), *Social work in the era of devolution: Toward a just practice* (pp. 351–369). New York: Fordham University Press.

Foley, W.A. (2005). Do humans have innate mental structures? Some arguments from linguistics. In S. McKinnon, & S. Silverman (Eds.), *Complexities: Beyond nature and nurture* (pp. 43–63). Chicago: University of Chicago Press.

Friibergh Workshop (2000). *Sustainability science: Statement of the Friibergh workshop on sustainability science.* Retrieved November 4, 2008, from ksgnotes1.harvard.edu/BCSIA/sust.nsf/pubs/pub3.

Frumkin, H., Frank, L., & Jackson, R. (2004). *Urban sprawl and public health: Designing, planning, and building for health communities.* Washington: Island Press.

Gale, F., Bolzan, N., & McRae-McMahon, D. (Eds.) (2007). *Spirited practices: Spirituality and the helping professions.* Sydney: Allen & Unwin.

Garber, K. (2007). Technology's morning after. *U.S. News & World Report* (December 20). Retrieved October 21, 2008, from www.usnews.com/articles/news/2007/12/20/technologys-morning-after.html.

Garvin, C.D., & Seabury, B.A. (1997). *Interpersonal practice in social work: Promoting competence and social justice* (2nd ed.). Boston: Allyn & Bacon.

Gauldie, S. (1969). *Architecture: The appreciation of the arts.* Oxford: Oxford University Press.

Geography Education National Implementation Project. (1986). *Maps, the landscape, and fundamental themes in geography.* Washington: National Geographic Society.

Gergen, K. (1985). The social constructionist movement in modern psychology. *American Psychologist, 40*(3), 266–275.

Germain, C.B. (1981). The physical environment and social work practice. In A.N. Maluccio (Ed.), *Promoting competence in clients: A new/old approach to social work practice* (pp. 103–124). New York: Free Press.

Germain, C.B., & Gitterman, A. (1980). *The life model of social work practice.* New York: Columbia University Press.

Germain, C.B., & Gitterman, A. (1987). Ecological perspective. In A. Minahan (Ed.), *Encyclopedia of social work* (18th ed.) (pp. 488–499). Silver Spring: National Association of Social Workers.

Gifford, R. (Managing Ed.) (2008). Description. *Journal of Environmental Psychology.* Retrieved August 13, 2008, from www.elsevier.com/wps/find/journaldescription.cws_home/622872/description#description.

Gilbert, M.C. (2000). Spirituality in social work groups: practitioners speak out. *Social Work with Groups, 22*(4), 67–84.

Gilgun, J.F. (2005). An ecosystemic approach to assessment. In B.R. Compton, B. Galaway, & B.R. Cournoyer (Eds.), *Social work processes* (7th ed.) (pp. 349–360). Belmont: Brooks/Cole.

Gingrich, L.G. (2003). Theorizing social exclusion: Determinants, mechanisms, dimensions, forms, and acts of resistance. In W. Shera (Ed.), *Emerging perspectives on anti-oppressive practice* (pp. 3–23). Toronto: Canadian Scholars' Press Inc.

Ginsberg, L.H. (1969). Education for social work practice in rural areas. *Social Work Education, 15*(1).

Ginsberg, L.H. (1971). Rural social work. In *Encyclopedia of social work* (16th issue, vol. 2) (pp. 1138–1144). Washington: NASW Press.

Ginsberg, L.H. (Ed.) (1976, 1994, 1998). *Social work in rural communities* (1st, 2nd, 3rd eds.). Alexandria: CSWE Press.

Ginsberg, L.H. (1998). Introduction: An overview of rural social work. In L.H. Ginsberg (Ed.), *Social work in rural communities* (3rd. ed.) (pp. 3–22). Alexandria: CSWE Press.

Gitterman, A. (2002) The life model. In A.R. Roberts, & G.J. Greene (Eds.), *Social workers' desk reference* (pp. 105–108). New York: Oxford University Press.

Gold, N. (2002). The nature and function of assessment. In F.J. Turner (Ed.), *Social work practice: A Canadian perspective* (2nd ed.) (pp. 143–154). Toronto: Pearson Education Canada.

Goode, J. (2002). The two cultures: How the rise of the brands is changing the face of wine. Retrieved May 16, 2007, from www.wineanorak.com/twocultures.htm.

Goode, J. (2003, September 12). Mechanisms of terroir. *Harpers Weekly.* Retrieved May 16, 2007, from www.wineanorak.com/mechanisms_terroir.htm.

Gordon, W.E. (1969). Basic concepts for an integrative and generative conception of social work. In G. Hearn (Ed.), *The general systems approach: Contributions towards an holistic conception of social work* (pp. 5–11). Alexandria: CSWE Press.

Gordon, W.E. (1981). A natural classification system for social work literature and knowledge. *Social Work, 26*(2), 134–138.

Gore, A. (2006). *An inconvenient truth: The planetary emergency of global warming and what we can do about it.* New York: Rodale.

Gottlieb, J., & Gottlieb, L. (2004). A visit to Dwight's Hollow. In R.F. Rivas, & G.H. Hull, Jr. (Eds.), *Case studies in generalist practice* (3rd ed.) (pp. 102–108). Belmont: Brooks/Cole.

Graham, J.R., Swift, K.J., & Delaney, R. (2009). *Canadian social policy: An introduction* (3rd ed.). Toronto: Pearson Education Canada.

Grange, J. (1977). On the way towards foundational ecology. *Soundings: An Interdisciplinary Journal, 60*(2), 135–149.

Graveline, F.J. (1998). *Circle works: Transforming Eurocentric consciousness.* Halifax: Fernwood Publishing.

Grinnell, R.M., Jr., Kyte, N.S., & Bostwick, G.J., Jr. (1981). Environmental modification. In A.N. Maluccio (Ed.), *Promoting competence in clients: A new/old approach to social work practice* (pp. 152–184). New York: Free Press.

Guevara-Stone, L. (2008). Viva la revolucion energetica. *Alternatives Journal, 34*(6), 22–24.

Gutheil, I.A. (1992). Considering the physical environment: An essential component of good practice. *Social Work, 37*(5), 391–396.

Haas, T., & Nachtigal, P. (1998). *Place value: An educator's guide to good literature on rural lifeways, environments, and purposes of education.*

Charleston, West Virginia: Clearinghouse on Rural Education and Small Schools (Appalachia Educational Laboratory).

Hall, A. (1996). Social work or working for change? Action for grassroots sustainable development in Amazonia. *International Social Work, 39,* 27–33.

Hall, P. (1996). *Cities of tomorrow.* Oxford: Blackwell.

Hall, S.S. (2004). I, Mercator. In K. Harmon (Ed.), *Your are here: Personal geographies and other maps of the imagination* (pp. 15–19). New York: Princeton Architectural Press.

Hamelin, L.E. (1984). Summary and comments of the rapporteur [Offprint from the special issue "Managing Canada's North: Challenges and Opportunities. 16th National Seminar]. *Canadian Public Administration, 27*(2), 165–181.

Hancock, M.R. (1997). *Principles of social work practice: A generic practice approach.* New York: Haworth Press.

Hancock, M.R., & Millar, K.I. (1993). *Cases for intervention planning: A source book.* Chicago: Nelson-Hall.

Hanson, M. (2002). Practice with organizations. In M.A. Mattaini, C.T. Lowery, & C.H. Meyer (Eds.), *Foundations of social work practice: A graduate text* (pp. 263–290). Washington: NASW Press.

Haraway, D. (1995). Otherworldly conversations, terran topics, local terms. In V. Shiva, & I. Moser (Eds.), *Biopolitics: A feminist and ecological reader on biotechnology* (pp. 69–92). London: Zed Books.

Hardcastle, D.A., Wenocur, S., & Powers, P.R. (1997). *Community practice: Theories and skills for social workers.* New York: Oxford University Press.

Hare, I. (2004). Defining social work for the 21st century: The International Federation of Social Workers' revised definition of social work. *International Social Work, 47*(3), 407–424.

Harmon, K. (Ed.) (2004). *You are here: Personal geographies and other maps of the imagination.* New York: Princeton Architectural Press.

Harper, E. (2004). What Canada means to me. In J. Oakes, R. Riewe, Y. Belanger, S. Blady, K. Legge, & P. Wiebe (Eds.), *Aboriginal cultural landscapes* (pp. 163–167). Winnipeg: Aboriginal Issues Press (University of Manitoba).

Hart, M.A. (1996). Sharing circles: Utilizing traditional practice methods for teaching, helping, and supporting. In S. O'Meara, & D.A. West (Eds.), *From our eyes: Learning from indigenous peoples* (pp. 59–72). Toronto: Garamond Press.

Hart, M.A. (2002). *Seeking mino-pimatisiwin: An Aboriginal approach to helping*. Halifax: Fernwood.

Hart, M.A. (2006). An Aboriginal approach to social work practice. In T. Heinonen, & L. Spearman (Eds.), *Social work practice: Problem solving and beyond* (2nd ed.) (pp. 235–259). Toronto: Nelson.

Hawley, A.H. (1986). *Human ecology: A theoretical essay*. Chicago: University of Chicago Press.

Healy, L. (2001). *International social work: Professional action in an interdependent world*. New York: Oxford University Press.

Heft, H. (2001). *Ecological psychology in context*. Mahwah: Lawrence Erlbaum.

Heinonen, T., & Spearman, L. (2006). *Social work practice: Problem solving and beyond* (2nd ed.). Toronto: Nelson.

Hepworth, D.H., Rooney, R.H., & Larsen, J.A. (1997). *Direct social work practice: Theory and skills* (5th ed.). Belmont: Brooks/Cole.

Hick, S. (2006). *Social work in Canada: An introduction* (2nd ed.). Toronto: Thompson Educational Publishing.

Higgins, L. (2002). Towards community music conceptualization. In H. Schippers, & N. Kors (Eds.), *Proceedings of the 2002 ISME (International Society for Music Education) Commission on Community Music Activity*. Rotterdam: ISME.

Hillier, J., & Rooksby, E. (Eds.) (2005). *Habitus: A sense of place* (2nd ed.). Aldershot: Ashgate.

Hodge, D.R. (2001). Spiritual assessment: A review of major qualitative methods and a new framework for assessing spirituality. *Social Work, 46*(3), 203–214.

Hodge, D.R., & Kaopua, L.S. (2005). Spiritual assessment: An overview of its importance and six instruments for conducting assessments. *Spirituality and Social Work Forum, 12*(1), 3–5.

Hoff, M.D., & McNutt, J.G. (Eds.) (1994). *The global environmental crisis: Implications for social welfare and social work*. Aldershot: Ashgate

Hoff, M.D., & Polack, R.J. (1993). Social dimensions of the environmental crisis: Challenges for social work. *Social Work, 38*(2), 204–211.

Hoffman, K.S., & Sallee, A.L. (1994). *Social work practice: Bridges to change.* Boston: Allyn & Bacon.

Hokenstad, J.S., Khinduka, S., & Midgley, J. (Eds.) (1992). *Profiles in international social work.* Washington: NASW Press.

Holmes, E., & Byrne-Armstrong, H. (2007). Aboriginal healing, dreaming, and Western medicine. In F. Gale, N. Bolzan, & D. McRae-McMahon (Eds.), *Spirited practices: spirituality and the helping professions* (pp. 43–50). Sydney: Allen & Unwin.

Holst, W. (1997). Aboriginal spirituality and environmental respect. *Social Compass, 44*(1), 145–156.

Hull, G.H., Jr., & Kirst-Ashman, K.K. (2004). *The generalist model of human services practice.* Belmont: Brooks/Cole.

Hutchison, D. (2004). *A natural history of place in education.* New York: Teachers College Press (Columbia University).

Ife, J. (2000). Localized needs and a globalized economy: Bridging the gap with social work practice [Special issue—social work and globalization]. *Canadian Social Work Review, 2*(1), 50–64.

Ingersoll, E. (2001). The spiritual wellness inventory. In C. Faiver, R.E. Ingersoll, E. O'Brien, & C. McNally (Eds.), *Explorations in counseling and spirituality: Philosophical, practical, and personal reflections* (pp. 185–194). Belmont: Wadsworth.

International Association of Schools of Social Work (IASSW). (2001, June). *Definition of social work.* Retrieved November 21, 2008, from www.iassw-aiets.org/index.php?option=com_content&task=blogcategory&id=26&Itemid=51.

International Association of Schools of Social Work (IASSW). (2008). *List of member schools.* Retrieved November 21, 2008, from www.iassw-aiets.org/index.php?option=com_content&task=blogcategory&id=34&Itemid=57.

International Council on Social Welfare (ICSW). (2008). *Our members.* Retrieved November 21, 2008, from www.icsw.org/intro/ourmembe.htm.

International Federation of Social Workers (IFSW). (2000, July). *Definition of social work*. Retrieved November 21, 2008, from www.ifsw.org/en/f38000138.html.

International Federation of Social Workers (IFSW). (2004). *International policy statement on globalisation and the environment*. Retrieved November 21, 2008, from www.ifsw.org/en/p38000222.html.

International Federation of Social Workers (IFSW). (2008). *Membership*. Retrieved November 21, 2008, from www.ifsw.org/en/p38000060.html.

Irwin, A. (2001). *Sociology and the environment*. Cambridge: Polity.

Jacobs, C. (1997). On spirituality and social work practice. *Smith College Studies in Social Work, 67*(2), 171–175.

Jacobs, D.T. (Four Arrows) (Ed.) (2006). *Unlearning the language of conquest: Scholars expose anti-Indianism in America*. Austin: University of Texas Press.

Jacobs, J. (2004). *Dark age ahead*. Toronto: Vintage Canada.

James, G.G. (2005). International social welfare. In J.C. Turner, & F.J. Turner (Eds.), *Canadian social welfare* (5th ed.) (pp. 485–502). Toronto: Pearson Education Canada.

Johnson, H.W. (Ed.) (1980). *Rural human services: A book of readings*. Itasca: Peacock.

Johnson, L.C. (1998). *Social work practice: A generalist approach* (6th ed.). Boston: Allyn & Bacon.

Johnson, L.C., & Yanca, S.J. (2007). *Social work practice: A generalist approach* (9th ed.). Boston: Allyn & Bacon.

Johnson, S.K. (1997). Does spirituality have a place in rural social work? *Social Work and Christianity, 24*(1), 58–66.

Karls, J.M. (2002). Person-in-environment system: Its essence and applications. In A.R. Roberts, & G.J. Greene (Eds.), *Social workers' desk reference* (pp. 194–198). New York: Oxford University Press.

Karls, J.M., & Wandrei, K.E. (1994). *Person-in-environment system: The PIE classification system for social functioning problems*. Washington: NASW Press.

Karls, J.M., & Wandrei, K.E. (2000). *CompuPIE: Interactive software for PIE*. Washington: NASW Press.

Kasiram, M.I. (1998). Achieving balance and wellness through spirituality: A family therapy perspective. *Social Work/Maatskaplike (South Africa), 34*(2), 171–175.

Kauffman, S.E. (1994). Citizen participation in environmental decisions: Policy, reality, and considerations for community organizing. In M.D. Hoff, & J.G. McNutt (Eds.), *The global environmental crisis: Implications for social welfare and social work* (pp. 219–239). Aldershot: Ashgate.

Kauffman, S.E., Walter, C.A., Nissly, J., & Walker, J. (1994). Putting the environment into the human behaviour and the social environment curriculum. In M.D. Hoff, & J.G. McNutt (Eds.), *The global environmental crisis: Implications for social welfare and social work* (pp. 277–296). Aldershot: Ashgate.

Kemmis, D. (1992). *Community and the politics of place*. Norman: University of Oklahoma Press.

Kimberley, M.D., & Osmond, M.L. (2005). The profession: An integrative perspective. In J.C Turner, & F.J. Turner (Eds.), *Canadian social welfare* (5th ed.) (pp. 404–425). Toronto: Pearson Education Canada.

Kirst-Ashman, K.K. (2007). *Introduction to social work and social welfare: Critical thinking perspectives* (2nd ed.). Belmont: Brooks/Cole.

Kirst-Ashman, K.K., & Hull, G.H., Jr. (1997). *Generalist practice with organizations and communities*. Chicago: Nelson-Hall.

Kluger, J. (2007, April 9). What now? Our feverish planet badly needs a cure. *TIME (Canadian Edition—Special Double Issue), 169*(15), 46–57.

Kolari, T. (2004). *The right to a decent environment with special reference to Indigenous peoples* (Research Report, Juridica Lapponica #31). Rovaniemi: Univesity of Lapland Arctic Centre (The Northern Institute for Environmental and Minority Law).

Kretzmann, J.P., & McKnight, J.L. (1993). *Building communities from the inside out: A path toward finding and mobilizing a community's assets*. Evanston: Northwestern University Center for Urban Affairs and Policy Research.

LaChappelle, D. (1988). *Sacred land, sacred sex, rapture of the deep: Concerning deep ecology and celebrating life*. Durango: Kivaki Press.

Laghi, B. (2007, January 26). Climate concerns now top security and health. *Globe & Mail*, pp. A1, A4.

Lane, B.C. (2001). *Landscapes of the sacred: Geography and narrative in American spirituality (expanded edition)*. Baltimore: Johns Hopkins University Press.

Latta, A. (2007). Environmental citizenship: A model linking ecology with social justice could lead to a more equitable future. *Alternatives: Canadian Environmental Ideas & Action, 33*(1), 18–19.

Leach, N. (2005). Belonging: Towards a theory of identification with space. In J. Hillier, & E. Rooksby (Eds.), *Habitus: A sense of place* (2nd ed.) (pp. 297–311). Aldershot: Ashgate.

LeCroy, C.W. (Ed.). (1999). *Case studies in social work practice* (2nd ed.). Belmont: Brooks/Cole.

Lee, E. (2004). Epilogue: Landscapes, perspectives, and nations—what does it all mean? In I. Krupnik, R. Mason, T. Horton (Eds.), *Northern ethnographic landscapes: Perspectives from circumpolar nations* (pp. 401–406). Washington: Arctic Studies Center (Smithsonian Institute).

Lehmann, P., & Coady, N. (2001). *Theoretical perspectives for direct social work practice: A generalist-eclectic approach*. New York: Springer.

Lerner, S. (Ed.) (1993). *Environmental stewardship: Studies in active earthkeeping*. Waterloo: University of Waterloo.

Lewis, G.H. (1992). Who do you love? The dimensions of musical taste. In J. Lull (Ed.), *Popular music and communication* (2nd ed.). London: Sage.

Lightman, E. (2003). *Social policy in Canada*. Don Mills: Oxford University Press.

Lindholm, K. (1988). In search of an identity—social work training in the Nordic countries [Special supplementary issue in English of *Nordisk Sosialt Arbeid* on the occasion of IFSW's World Conference in Stockholm, 1988]. *Nordic Journal of Social Work, 8*, 4–14.

Little Bear, L. (2000). Foreword. In G. Cajete, *Native science: Natural laws of interdependence*. Santa Fe: Clear Light Publishers.

Locke, B., Garrison, R., & Winship, L. (1998). *Generalist social work practice: Context, story, and partnerships.* Belmont: Brooks/Cole.

Lohman, N. (2005). Social work education for rural practice. In N. Lohman, & R.A. Lohman (Eds.), *Rural social work practice* (pp. 293–312). New York: Columbia University Press.

Longclaws, L. (2005). Social work and the medicine wheel framework. In B.R. Compton, B. Galaway, & B.R. Cournoyer (Eds.), *Social work processes* (7th ed.) (pp. 361–367). Belmont: Brooks/Cole.

Lovell, M.L., & Johnson, D.L. (1994). The environmental crisis and direct social work practice. In M.D. Hoff, & J.G. McNutt (Eds.), *The global environmental crisis: Implications for social welfare and social work* (pp. 199–218). Aldershot: Ashgate.

Lovelock, J. (1987). *Gaia.* Oxford: Oxford University Press.

Lundy, C. (2004). *Social work and social justice: A structural approach to practice.* Peterborough: Broadview Press.

Lysack, M. (2003). "When the sacred shows through": Narratives and reflecting teams in counsellor education. *Sciences Pastorals/Pastoral Sciences, 22*(1), 115–146.

Lysack, M. (2007). Family therapy, the ecological self, and global warming. *Context, 91*(June), 9–11.

Lysack, M. (2008). The influence of the built environment on the emotional health of children and families. Presentation to the Southern Alberta Child & Youth Health Network, February 27, Calgary.

Mabogunje, A.L. (1996). Introduction [Special issue on geography: state of the art I—the environmental dimension]. *International Social Science Journal, 150,* 443–447.

Mace, W.M. (Ed.) (2008). Instructions for authors. *Ecological Psychology Journal.* Retrieved August 13, 2008, from www.trincoll.edu/depts/ecopsyc/isep/journal.html.

Maclouf, P., & Lion, A. (1984). *Aging in remote rural areas: A challenge to social and medical service.* Vienna: European Centre for Social Welfare Training and Research.

MacQueen, K. (2007). The trials of Saint David. *Maclean's, 120*(43), 66–73.

Mallett, K., Bent, K., & Josephson, W. (2000). Aboriginal Ganootamaage justice services of Winnipeg. In J. Proulx, & S. Perrault (Eds.), *No place for violence: Canadian Aboriginal alternatives* (pp. 58–77). Halifax: Fernwood and RESOLVE (Research and Education for solutions to Violence and Abuse).

Maluccio, A.N., Washitz, S., & Libassi, M.F. (1999). Ecologically oriented, competence-centered social work practice. In C.W. LeCroy (Ed.), *Case studies in social work practice* (2nd ed.) (pp. 31–38). Belmont: Brooks/Cole.

Marlow, C., & Van Rooyen, C. (2001). How green is the environment in social work? *International Social Work, 44*(2), 241–254.

Martin, J. (2006). *The meaning of the 21st century: A vital blueprint for ensuring our future.* New York: Riverhead Books.

Martinez-Brawley, E.E. (1981). Rural social and community work as political movements in the United States and United Kingdom. *Community Development Journal, 16*(3), 201–211.

Martinez-Brawley, E.E. (1990). *Perspectives on the small community: Humanistic views for practitioners.* Washington: NASW Press.

Mathews, F. (1991). *The ecological self.* London: Routledge.

Mathews, F. (2005). *Reinhabiting a reality: Towards a recovery of culture.* Albany: SUNY Press.

Mattaini, M.A. (2002). Practice with individuals. In M.A. Mattaini, C.T. Lowery, & C.H. Meyer (Eds.), *Foundations of social work practice: A graduate text* (pp. 151–183). Washington: NASW Press.

Mattaini, M.A., Lowery, C.T., & Meyer, C.H. (Eds.) (2002). *Foundations of social work practice: A graduate text.* Washington: NASW Press.

Matthies, A.L., Jarvela, M., & Ward, D. (2000). An eco-social approach to tackling exclusion in European cities: A comparative research project in progress. *European Journal of Social Work, 3*(1), 43–51.

McCormack, P. (1998). Native homelands as cultural landscape: Decentering the wilderness paradigm. In J. Oakes, R. Riewe, K. Kinew, & E. Maloney (Eds.), *Sacred lands: Aboriginal world views, claims, and conflicts* (pp. 25–32). Edmonton: Canadian Circumpolar Institute (University of Alberta).

McGaa, E. (1990). *Mother Earth spirituality: Native American paths to healing ourselves and our world*. San Francisco: HarperSanFrancisco.

McGee, H., & Patterson, D. (2007, May 6). Talk dirt to me. *New York Times Living Magazine (final edition)*, p. 76.

McKay, S. (1987). Social work in Canada's north: Survival and development issues affecting aboriginal and industry-based economies. *International Social Work, 30*, 259–278.

McKay, S. (2002). Postmodernism, social well-being, and the mainstream/progressive debate. In F.J. Turner (Ed.), *Social work practice: A Canadian perspective* (2nd ed.) (pp. 20–32). Toronto: Pearson Education Canada.

McKenzie, P. (2001). Aging people in aging places: Addressing the needs of older adults in rural Saskatchewan. *Rural Social Work, 6*(3), 74–83.

McKinnon, J. (2001). Social work and the environment. In M. Alston, & J. McKinnon (Eds.), *Social work: Fields of practice* (pp. 193–205). Melbourne: Oxford University Press.

McKinnon, J. (2005). Social work, sustainability, and the environment. In M. Alston, & J. McKinnon (Eds.), *Social work: Fields of practice* (2nd ed.) (pp. 225–236). Melbourne: Oxford University Press.

McSwan, D., & McShane, M. (Eds.) (1994). *Proceedings of the International Conference on Issues Affecting Rural Communities* (Townsville, Australia, July). Townsville: Rural Education Research and Development Centre of James Cook University.

Meadowcroft, J. (2007). Building the environmental state: What the history of social welfare tells us about the future of environmental policy. *Alternatives: Canadian Environmental Ideas & Action, 33*(1), 11–17.

Meawasige, I. (1995). The healing circle. In R. Delaney, & K. Brownlee (Eds.), *Northern social work practice* (pp. 136–146). Thunder Bay: Lakehead University Centre for Northern Studies.

Mermelstein, J., & Sundet, P.A. (1998). Rural social work is an anachronism: The perspective of twenty years of experience and debate. In L.H. Ginsberg (Ed.), *Social work in rural communities* (3rd ed.) (pp. 63–80). Alexandria: CSWE Press.

Meyer, C.H. (1976). *Social work practice* (2nd ed.). New York: Free Press.

Meyer, C.H. (1995). The ecosystems perspective: Implications for practice. In C.H. Meyer, & M.A. Mattaini (Ed.s), *The foundations of social work practice: A graduate text* (pp 16–27). Washington: NASW Press.

Midgley, J. (1981). *Professional imperialism: Social work in the Third World.* London: Heinemann.

Midgley, J. (1997). *Social welfare in global context.* Thousand Oaks: Sage Publications.

Miley, K.K., O'Melia, M., & DuBois, B. (2004). *Generalist social work practice: An empowering approach* (4th ed.). Boston: Pearson Education.

Mitchell, J. (1990). Human dimensions of environmental hazards. In A. Kirby (Ed.), *Nothing to fear: Risks and hazards in American society* (pp. 131–175). Tucson: University of Arizona Press.

Mittelstaedt, M., Galloway, G., & Laghi, B. (2007, February 1). Harper puts green machine in motion. *Globe & Mail*, pp. A1, A4.

Mokuau, N., & Iuli, B. P. (2004). Nalani Ethel: Social work with a Hawaiian woman and her family. In R.F. Rivas, & G.H. Hull, Jr. (Eds.), *Case studies in generalist practice* (3rd ed.) (pp. 22–28). Belmont: Brooks/Cole.

Morales, A.T., & Sheafor, B.N. (2004). *Social work: A profession of many faces* (10th ed.). Boston: Pearson Education.

Morito, B. (2002). *Thinking ecologically: Environmental thought, values and policy.* Halifax: Fernwood.

Morrissette, V., McKenzie, B., & Morrissette, L. (1993). Towards an aboriginal model of social work practice: Cultural knowledge and traditional practices. *Canadian Social Work Review, 10*(1), 91–108.

Mullaly, B. (2002). *Challenging oppression: A critical social work approach.* Don Mills: Oxford University Press.

Mullaly, B. (2006). *The new structural social work: Ideology, theory, practice* (3rd ed.). Don Mills: Oxford University Press.

Naess, A. (1973). The shallow and the deep, long range ecology movement. *Inquiry, 16*(2), 95–100.

Naidoo, K. (2004). "It's time to act." *Canadian Review of Social Policy, 53*, 11–17.

Narhi, K. (2004). *The eco-social approach in social work and the challenges to expertise in social work.* Jyvaskyla: University of Jyvaskyla.

National Association of Social Workers (NASW). (2000). *Social work speaks: NASW policy statements.* Washington: NASW Press.

National Research Council. (1997). *Rediscovering geography: New relevance for science and society.* Washington: National Academy Press.

Neil, R., & Smith, M. (1998). Education and sacred land: First Nations, Metis, and Taoist views. In J. Oakes, R. Riewe, K. Kinew, & E. Maloney (Eds.), *Sacred lands: Aboriginal world views, claims, and conflicts* (pp. 87–100). Edmonton: Canadian Circumpolar Institute (University of Alberta).

Nelson, C., & McPherson, D. (2004). Contextual fluidity: An emerging practice model for helping. *Rural Social Work, 9,* 199–208.

Nelson, C., McPherson, D., & Kelley, M.L. (1987). Contextual patterning: A key to human service effectiveness in the North. In P. Adams, & D. Parker (Eds.), *Canada's subarctic universities* (pp. 66–82). Ottawa: Association of Canadian Universities for Northern Studies.

Netting, F.E., Kettner, P.M., & McMurtry, S.L. (1993). *Social work macro practice.* New York: Longman.

Neugeboren, B. (1996). *Environmental practice in the human services: Integration of micro and macro roles, skills, and contexts.* New York: Haworth Press.

Neuman, K. (2007). Topping the chart: Canadian priorities. *Alternatives: Canadian Environmental Ideas & Action, 33*(1), 15.

Newby, H. (1991). One world, two cultures: Sociology and the environment. *Network, 50,* 1–8.

Noble, C. (2004). Postmodern thinking: Where is it taking social work? *Journal of Social Work, 4*(3), 289–304.

Norberg-Shulz, C. (1980). *Genius loci: Towards a phenomenology of architecture.* New York: Rizzoli.

Norton, W. (2004). *Human geography* (5th ed.). Don Mills: Oxford University Press.

O'Neill, J., & Horner, W. (1984). Rural insights from theory and research. *Human Services in the Rural Environment, 9*(1), 2–3.

Orr, D.W. (1992). *Ecological literacy: Education and the transition to a postmodern world.* Albany: SUNY Press.

Orr, D.W. (1994). *Earth in mind: On education, environment and the human prospect*. Washington: Island Press.

Orr, D.W. (2004). The learning curve [Special issue on ecoliteracy]. *Resurgence, 226*(September/October). Available at www.resurgence.org/magazine/issue226-Ecoliteracy-Dancing-Earth.html.

Ortega y Gasset, J. (1985). *Meditations on hunting*. New York: Scribners.

Ortiz, L., & Smith, G. (1999). The role of spirituality in empowerment practice. In W. Shera, & L.M. Wells (Eds.), *Empowerment practice in social work: Developing richer conceptual foundations* (pp. 307–319). Toronto: Canadian Scholars' Press Inc.

Pandey, S. (1998). Women, environment, and sustainable development. *International Social Work, 41*, 339–355.

Patterson, S., Jess, J., & LeCroy, C.W. (1999). Using the ecological model in generalist practice: Life transitions in late adulthood. In C.W. LeCroy (Ed.), *Case studies in social work practice* (2nd ed.) (pp. 10–14). Belmont: Brooks/Cole.

Payne, M. (2005). *Modern social work theory* (3rd ed.). Chicago: Lyceum.

Pearce, J. (1991). Ecological sensibility in recent Canadian children's novels. In R. Lorimer, M. M'Gonigle, J.P. Reveret, & S. Ross (Eds.), *To see ourselves/to save ourselves: Ecology and culture in Canada* (pp. 115–123). Montreal: Association for Canadian Studies.

Peat, F.D. (1994). *Lighting the seventh fire: The spiritual ways, healing, and science of the Native American*. New York: Birch Lane Press.

Penton, K. (1993). Ideology, social work, and the Gaian connection. *Australian Social Work, 46*(4), 41–48.

Perkins, L.B. (2001). *Building type basics for elementary and secondary schools*. New York: John Wiley & Sons.

Pile, S. (1997). Human agency and human geography revisited: a critique of "new models" of the self. In T. Barnes, & D. Gregory (Eds.), *Reading human geography: The poetics and politics of inquiry* (pp. 407–434). London: Arnold.

Pincus, A., & Minahan, A. (1973). *Social work practice: Model and method*. Itasca: Peacock.

Poulin, J. (2005). *Strengths-based generalist practice: A collaborative approach* (2nd ed.). Belmont: Brooks/Cole.

Prefontaine, A. (CEO and publisher). (2007). *Canadian environment awards 2007: A celebration of community achievement.* Ottawa: Canadian Geographic Enterprises.

Proshansky, H.M., Ittelson, W.H., & Rivlin, L.G. (1970). *Environmental psychology: Man and his physical setting.* New York: Holt, Rinehart, & Winston.

Proulx, J., & Perrault, S. (Eds.) (2000). *No place for violence: Canadian Aboriginal alternatives.* Halifax: Fernwood and RESOLVE (Research and Education for Solutions to Violence and Abuse).

Quieta, R. (2003). Community development and social work in the Philippines: Theory and practice. *Social Development Issues, 25*(3), 62–73.

Rajotte, F. (1998). *First Nations faith and ecology.* Toronto: Anglican Book Centre & United Church Publishing House.

Redclift, M., & Benton, T. (Eds.) (1994). *Social theory and the global environment.* New York: Routledge.

Relph, E. (1976). *Place and placelessness.* London: Pion.

Resnick, H., & Jaffee, B. (1982). The physical environment and social welfare. *Social Casework, 63*(6), 354–362.

Ribes, B. (1985). *Youth and life in remote rural areas.* Vienna: European Centre for Social Welfare Training and Research.

Rice, S. (2002). Magic happens: Revisiting the spirituality and social work debate. *Australian Social Work, 55*(4), 303–312.

Richmond, M. (1922). *What is social work?* New York: Russell Sage Foundation.

Rivas, R.F., & Hull, G.H., Jr. (2004). *Case studies in generalist practice* (3rd ed.). Belmont: Brooks/Cole.

Rodwell, M.K. (1995). After democracy breaks out: Challenges in the preparation of social workers for new roles in Brazilian society. In G. Rogers (Ed.), *Social work field education: Views and Visions.* Dubuque: Kendall/Hunt.

Rogge, M.E. (1994). Environmental injustice: Social welfare and toxic waste. In M.D. Hoff, & J.G. McNutt (Eds.), *The global environmental crisis: Implications for social welfare and social work* (pp. 53–74). Aldershot: Ashgate.

Rogge, M.E., & Cox, M.E. (2001). The person-in-environment perspective in social work journals: A computer-assisted content analysis. *Journal of Social Service Research, 28*(2), 47–68.

Rosenhek, R. (2006). Earth, spirit, and action: The deep ecology movement as spiritual engagement. *The Trumpeter: Journal of Ecosophy, 22*(2), 90–95.

Rosenman, L.S. (1980). Social work education in Australia: The impact of the American model. *Journal of Education for Social Work, 16*(1), 112–118.

Roszak, T. (1972). *Where the wasteland ends.* Garden City: Doubleday.

Rothery, M. (2002). The resources of intervention. In F.J. Turner (Ed.), *Social work practice: A Canadian perspective* (2nd ed.) (pp. 241–254). Toronto: Pearson Education Canada.

Rothery, M. (2008). Critical ecological systems theory. In N. Coady, & P. Lehmann (Eds.), *Theoretical perspectives for direct social work practice: A generalist-eclectic approach* (2nd ed.) (pp. 89–118). New York: Springer Publishing.

Rowe, B. (2005). Contemporary issues in biopsychosocial functioning. In J.C. Turner, & F.J. Turner (Eds.), *Canadian social welfare* (5th ed.) (pp. 196–207). Toronto: Pearson Education Canada.

Rubenstein, J.M. (2005). *An introduction to human geography: The cultural landscape* (8th ed.). Upper Saddle River: Pearson Prentice Hall.

Russel, R. (1998). Spirituality and religion in graduate social work education. In E.R. Canda (Ed.), *Spirituality in social work: New directions* (pp. 15–30). New York: Haworth Press.

Sahtouris, E. (1992). The survival path: Cooperation between indigenous and industrial humanity. *Proceedings of the United Nations Policy Meeting on Indigenous Peoples.* Santiago, Chile. Retrieved September 8, 2008, from www.ratical.org/LifeWeb/Articles/survival.html.

Sainte-Marie, B. (1992). Preface. In A.D. Manitopes, & O. Courchene (Eds.), *Voice of the eagle: The final warning message of the Indigenous people of*

Turtle Island presented to the people of Mother Earth (Earth Summit, Rio de Janiero, June). Calgary: Aboriginal Awareness Society.

Saleebey, D. (2004). The power of place: Another look at the environment. *Families in Society, 85*(1), 7–16.

Saleebey, D. (2007). *The strengths perspective in social work practice* (4th ed.). Boston: Allyn & Bacon.

Scales, T.L., & Streeter, C.L. (2004a). Asset building to sustain rural communities. In T.L. Scales, & C.L. Streeter (Eds.), *Rural social work: Building and sustaining community assets* (pp. 1–6). Belmont: Brooks/Cole.

Scales, T.L., & Streeter, C.L. (Eds.) (2004b). *Rural social work: Building and sustaining community assets*. Belmont: Brooks/Cole.

Schmidt, G. (2005). Geographic context and northern child welfare practice. In K. Brownlee, & J.R. Graham (Eds.), *Violence in the family: Social work readings and research from northern and rural Canada* (pp. 16–29). Toronto: Canadian Scholars' Press Inc.

Schmidt, G., & Klein, R. (2004). Geography and social worker retention. *Rural Social Work, 9*, 235–243.

Schobert, F.M, Jr., & Barron, D.A. (2004). Community development in an international setting: The role of sustainable agriculture in social work practice. In T.L. Scales, & C.S. Streeter (Eds.), *Rural social work: Building and sustaining community assets* (pp. 178–191). Belmont: Brooks/Cole.

Scott, B.A., & Kroger, S.M. (2006). *Teaching psychology for sustainability: A manual of resources*. Retrieved April 28, 2007, from www.teachgreenpsych.com.

Seed, J., Macy, J., Fleming, P., & Naess, A. (1988). *Thinking like a mountain: Towards a council of all beings*. Gabriola Island: New Society Publishers.

Sermabeikian, P. (1994). Our clients, ourselves: The spiritual perspective and social work practice. *Social Work, 39*(2), 178–183.

Sheafor, B.W., & Horejsi, C.R. (2008). *Techniques and guidelines for social work practice* (8th ed.). Boston: Pearson Education.

Shelemay, K.K. (2001) *Soundscapes: Exploring music in a changing world*. New York: Norton.

Sheridan, M.J. (2007). Earth as a source of spirit. In K. van Wormer, F.H. Besthorn, & T. Keefe (Eds.), *Human behavior and the social environment: Macro level* (pp. 269–272). New York: Oxford University Press.

Shubert, J.G. (1994). Case studies in community organizing around environmental threats. In M.D. Hoff, & J.G. McNutt (Eds.), *The global environmental crisis: Implications for social welfare and social work* (pp. 240–257). Aldershot: Ashgate.

Shulman, L. (2006). *The skills of helping: Individuals, families, groups, and communities* (5th ed.). Belmont: Brooks/Cole.

Smith, G. (2007). Grounding learning in place. *WorldWatch* (March/April), 20–24.

Soine, L. (1987). Expanding the environment in social work: The case for including environmental hazards content. *Journal of Social Work Education, 23*(2), 40–46.

Sowers, K.M., & Rowe, W.S. (2007). *Social work practice and social justice: From local to global perspectives.* Belmont: Brooks/Cole.

Spretnak, C. (1991). *States of grace: The recovery of meaning in the postmodern age.* New York: Harper Collins.

Stairs, A., & Wenzel, G. (1992). "I am I and the Environment": Inuit hunting, community, and identity. *Journal of Indigenous Studies, 3*(1), 1–12.

Strahler, A., & Strahler, A. (2005). *Physical geography: Science and systems of the human environment (Canadian version)* (3rd ed.). Hoboken: John Wiley & Sons.

Stuart, P.H. (2004). Social welfare and rural people: From the colonial era to the present. In T.L. Scales, & C.L. Streeter (Eds.), *Rural social work: Building and sustaining community assets* (pp. 21–33). Belmont: Brooks/Cole.

Suopajarvi, L. (1998). *Regional identity in Finnish Lapland.* Paper presented at the Third International Congress of Arctic Social Sciences, Copenhagen, Denmark.

Suzuki, D. (1999, June 14). Saving the Earth (Essays on the Millenium series). *Maclean's, 112*(24), 42–45.

Suzuki, D. (2002). *The sacred balance: Rediscovering our place in nature* (with A. McConnell). Vancouver: Greystone Books.

Taaffe, E.J., Gauthier, H.L., & O'Kelly, M.E. (1996). *Geography of transportation.* Upper Saddle River: Prentice-Hall.

Tall, D. (1996). Dwelling: Making peace with space and place. In W. Vitek, & W. Jackson, (Eds.), *Rooted in the land: Essays on community and place.* New Haven: Yale University Press.

Taylor, N. (1999). Town planning 'social,' not just 'physical'? In C.H. Greed (Ed.), *Social town planning* (pp. 29–43). London: Routledge.

Tester, F. (1994). In an age of ecology: Limits to voluntarism and traditional theory in social work practice. In M. Hoff, & J. McNutt (Eds.), *The global environmental crises: Implications for social work and social welfare* (pp. 240–257). Aldershot: Avebury.

Tester, F. (1997). *From the ground up: Community development as an environmental movement.* In B. Wharf, & M. Clague (Eds.), *Community organizing: Canadian experiences* (pp. 228–247). Don Mills: Oxford University Press.

Tigges, L.M. (Ed.) (2006). Special issue on community cohesion and place attachment. *American Behavioral Scientist, 50*(2), 139–275.

Timpson, J., & Semple, D. (1997). Bring home payahtakenemowin (peace of mind): Creating self-governing community services. *Native Social Work Journal, 1*(1), 87–101.

Tobin, M., & Walmsley, C. (Eds.) (1992). *Northern perspectives: Practice and education in social work.* Winnipeg: Manitoba Association of Social Workers and the University of Manitoba Faculty of Social Work.

Tolliver, W.F. (1997). Invoking the spirit: A model for incorporating the spiritual dimension of human functioning into social work practice. *Smith College Studies in Social Work, 67*(3), 477–486.

Tolson, E.R., Reid, W.J., & Garvin, C.D. (1994). *Generalist practice: A task-centered approach.* New York: Columbia University Press.

Torjman, S.R. (1984). Editorial. *The Social Worker, 52* (1), 3.

Truss, L. (2003). *Eats, shoots & leaves: The zero tolerance approach to punctuation.* New York: Gotham Books.

Tuan, Y.F. (1974). Space and place: Humanistic perspectives. *Progress in Geography, 6,* 211–252.

Turner, F.J. (2005a). Assessment. In F.J. Turner (Ed.), *Encyclopedia of Canadian social work* (p. 17). Waterloo: Wilfrid Laurier University Press.

Turner, F.J. (2005b). Canadian social welfare: A shared patrimony. In J.C. Turner, & F.J. Turner (Eds.), *Canadian social welfare* (5th ed.) (pp. 1–12). Toronto: Pearson Education Canada.

Turner, F.J. (Ed.) (2005c). *Encyclopedia of Canadian social work*. Waterloo: Wilfrid Laurier University Press.

Turner, F.J. (2005d). International practice. In F.J. Turner (Ed.), *Encyclopedia of Canadian social work* (p. 198). Waterloo: Wilfrid Laurier University Press.

Turner, J.C., & Turner, F.J. (Eds.) (1981). *Canadian social welfare* (1st ed.). Toronto: Collier Macmillan Canada.

Ungar, M. (2002). A deeper, more social ecological social work practice. *Social Service Review, 76*(3), 480–497.

Ungar, M. (2003). The professional social ecologist: Social work redefined. *Canadian Social Work Review, 20*(1), 5–23.

Ungar, M. (2004). Surviving as a postmodern social worker: Two Ps and three Rs of direct practice. *Social Work, 49*(3), 488–496.

Union of Concerned Scientists. (1992). *World scientists' warning to humanity*. Cambridge, MA: Union of Concerned Scientists. Available at www.ucsusa.org/about/1992-world-scientists.html.

United Nations World Commission on Environment and Development (Brundtlant Commission). (1987). *Our common future*. New York: Oxford University Press.

Van den Berg, J.H. (1961). *The changing nature of man*. New York: Delta.

VandenBos, G.R. (Ed.) (2007). *APA dictionary of psychology*. Washington: American Psychological Association Press.

van Wormer, K. (1997). *Social welfare: A world view*. Chicago: Nelson-Hall.

van Wormer, K., Besthorn, F.H., & Keefe, T. (2007). *Human behavior and the social environment: Macro level—groups, communities, and organizations*. New York: Oxford University Press.

Varley, A. (1994). The exceptional and the everyday: Vulnerability analysis in the international decade for natural disaster reduction. In A. Varley

(Ed.), *Disasters, development, and environment* (pp. 1–11). West Sussex: John Wiley & Sons.

Vitek, W., & Jackson, W. (Eds.) (1996). *Rooted in the land: Essays on community and place.* New Haven: Yale University Press.

Wackernagel, M., & Rees, W. (1996). *Our ecological footprint: Reducing human impact on the Earth.* Gabriola Island: New Society.

Wahlberg, J. (2004). Personal growth and self-esteem through cultural spiritualism: A Native American experience. In R.F. Rivas, & G.H. Hull, Jr. (Eds.), *Case studies in generalist practice* (3rd ed.) (pp. 65–73). Belmont: Brooks/Cole.

Wakefield, J.C. (1996). Does social work need the eco-systems perspective? Part 1: Is the perspective clinically useful? *Social Service Review, 70*(1), 1–32.

Walker, D.E. (1998). *Sacred geography in northwestern North America.* Retrieved September 8, 2008, from www.indigenouspeople.net/ipl_final.html.

Walmsley, C. (2005). *Protecting Aboriginal Children.* Vancouver: UBC Press.

Warner Music Canada. (2004). *A sense of place: music from and inspired by the people and places of Canada* (CD No. WTVD61650). Toronto: Warner Music Canada.

Warren, K.J. (1998). The power and the promise of ecological feminism. In S.J. Armstrong, & R.G. Botzler (Eds.), *Environmental ethics: Divergence and convergence* (2nd ed.) (pp. 471–480). New York: McGraw.

Watkins, T.R. (2004). Natural helping networks: Assets for rural communities. In T.L. Scales, & C.L. Streeter (Eds.), *Rural social work: Building and sustaining community assets* (pp. 65–76). Belmont: Brooks/Cole.

Webb, P. (2004). Interrogating the production of sound and place: The British phenomenon, from lunatic fringe to worldwide massive. In S. Whiteley, A. Bennett, & S. Hawkins (Eds.), *Music, space and place: Popular music and cultural identity* (pp. 66–85). Aldershot: Ashgate.

Weick, A. (1981). Reframing the person-in-environment perspective. *Social Work, 26*(2), 140–145.

Wenders, W. (2001). *A sense of place* (talk at Princeton University). Retrieved February 26, 2007, from www.wim-wenders.com/news_reel/2001/0103princeton.htm.

Wenzel, G.W. (1999). Traditional ecological knowledge and Inuit: Reflections on TEK research and ethics. *Arctic,* 52(2), 113–124.

Wharf, B. (1985). Toward a leadership role in human services: The case for rural communities. *The Social Worker,* 53(1), 14–20.

Wharf, B. (Ed.) (1991). [Special theme issue on social work in the North.] *The Northern Review,* 7.

Wharf, B., & McKenzie, B. (2004). *Connecting policy to practice in the human services* (2nd ed.). Don Mills: Oxford University Press.

Whiteley, S. (2004). Introduction. In S. Whiteley, A. Bennet, & S. Hawkins, (Eds.), *Music, space and place: Popular music and cultural identity* (pp. 1–22). Aldershot: Ashgate.

Wiggins Frame, M. (2003). *Integrating religion and spirituality into counseling: A comprehensive approach.* Belmont: Brooks/Cole.

Wikipedia, the free encyclopedia (2007). Smoke and mirrors. Retrieved January 30, 2007, from en.wikipedia.org/wiki/Smoke_and_mirrors.

Wint, E. (2000). Factors encouraging the growth of sustainable communities: A Jamaican study. *Journal of Sociology and Social Welfare,* 27(3), 119–133.

Winkler, I., & Chartoff, R. (Executive Producers), & Stallone, S. (Writer/ Director). (2006). *Rocky Balboa* [Motion picture]. United States: Metro-Goldwyn-Mayer Pictures/Columbia Pictures/Revolution Studios.

Witherspoon, G. (1977). *Language and art in the Navajo universe.* Ann Arbor: University of Michigan Press.

Yelaja, S.A. (1985). Concepts of social work practice. In S.A. Yelaja (Ed.), *An introduction to social work practice in Canada.* Scarborough: Prentice Hall Canada.

Zalitack, J. (2005, February 7). Over to you: I can say whatever I want. *Maclean's,* 118(6), 49.

Zapf, M.K. (1985). *Rural social work and its application to the Canadian north as a practice setting.* Toronto: University of Toronto, Faculty of Social Work.

Zapf, M.K. (1999). Location and knowledge-building: Exploring the fit of western social work with traditional knowledge. *Native Social Work Journal, 2*(1), 139–153.

Zapf, M.K. (2001). Notions of rurality. *Rural Social Work, 6*(3), 12–27.

Zapf, M.K. (2002). Geography and Canadian social work practice. In F.J. Turner (Ed.), *Social work practice: A Canadian perspective* (2nd ed.) (pp. 69–83). Toronto: Pearson Education Canada.

Zapf, M.K. (2005a). Remote practice. In F.J. Turner (Ed.), *Encyclopedia of Canadian social work* (pp. 322–323). Waterloo: Wilfrid Laurier University Press.

Zapf, M.K. (2005b). The geographic base of Canadian social welfare. In J.C. Turner, & F.J. Turner (Eds.), *Canadian social welfare* (5th ed.) (pp. 60–74). Toronto: Pearson Education Canada.

Zapf, M.K. (2005c). The spiritual dimension of person and environment: Perspectives from social work and traditional knowledge. *International Social Work, 48*(5), 633–642.

Zapf, M.K. (2007). Profound connections between person and place: Exploring location, spirituality, and social work. In J. Coates, J.R. Graham, B. Swartzentruber, & B. Ouellette (Eds.), *Spirituality and social work: Selected Canadian readings.* (pp. 229–242). Toronto: Canadian Scholars' Press Inc.

Zapf, M.K. (2008). Transforming social work's understanding of person and environment: Spirituality and the "common ground." *Journal of Religion & Spirituality in Social Work, 27*(1–2), 171–181.

Zapf, M.K., & Rogers, G. (2006, June). *People and place: An alliance between social work and environmental design.* Paper presented at the Canadian Association of Schools of Social Work Annual Conference, York University, Toronto.

Zastrow, C. (2004). *Introduction to social work and social welfare: Empowering people* (8th ed.). Belmont: Brooks/Cole.

Zastrow, C. (2007). *The practice of social work: A comprehensive worktext* (8th ed.). Belmont: Brooks/Cole.